SpringerBriefs in Biology

More information about this series at http://www.springer.com/series/10121

Olav Giere

Perspectives in Meiobenthology

Reviews, Reflections and Conclusions

 Springer

Olav Giere
Universität Hamburg (Emeritus)
Hamburg, Germany

ISSN 2192-2179 ISSN 2192-2187 (electronic)
SpringerBriefs in Biology
ISBN 978-3-030-13965-0 ISBN 978-3-030-13966-7 (eBook)
https://doi.org/10.1007/978-3-030-13966-7

Library of Congress Control Number: 2019932689

This Springer imprint is published by the registered company Springer Nature Switzerland AG
The registered company address is: Gewerbestrasse 11, 6330 Cham, Switzerland

Preface and Acknowledgements

The idea of writing this contribution developed when reading the keynote presentations in the Abstract Book of the 16th International Meiofauna Conference 2016. Especially those by our colleagues Richard M. Warwick, Michaela Schratzberger and Jeroen Ingels elicited and confirmed many thoughts and suggestions I had reflected on and discussed before. Thankful for these stimuli, I felt encouraged to write them down in a structured form. The text has been partly looked over by my colleagues Jon Norenburg (Washington DC, USA), Richard M. Warwick (Plymouth, England) and Walter Traunspurger (Bielefeld, Germany), and their support is much acknowledged. I also thank all those colleagues who sent me reprints or sometimes even yet unpublished manuscripts. Finally, the continuous, 'author-friendly' and non-bureaucratic support of the editors of *SpringerBriefs in Biology* within the Springer Nature Group (Heidelberg) is thankfully acknowledged.

Hamburg, Germany   Olav Giere

Contents

List of Figures

Chapter 1
Introduction

Intentions – A subjective attempt – The meiobenthos dilemma: triple negations – Out of the niche of disregard – Directions and structures of future meiobenthology – Advantages of work with meiofauna – Disseminating progress – Relevance of general overviews exemplified

This treatise is a subjective attempt to summarize thoughts on the future of meiobenthology. Based on recent publications in the field, it tries to reflect the state of the art, indicate gaps and risks, and outline promising perspectives and potentials in meiobenthology. Therefore, the papers mentioned are examples demonstrating the lines of reasoning and the reference lists are selective without claiming completeness. These papers will be characterized briefly, though usually not commented in detail. In order to make each chapter a fairly independent and separately readable unit, a certain overlap in their contents and references listed is deliberate.

Whoever is studying meiobenthic animals will face the 'triple negative-problem': 'Meiofauna—not dangerous, not delicious, not deleterious!' How then to attract interest in organisms that are below the size range of our innate perception and that lack any apparent appeal? At the same time, meiobenthic animals occur almost everywhere in significant numbers, and their regular presence in extreme environments is characteristic. They are closely integrated in complex food webs, have large ecosystemic relevance in both marine and freshwater habitats, and can provide considerable ecosystem services, as documented in recent reviews covering various meiobenthic topics of general importance (e.g. Schmidt-Araya et al. 2016; Schratzberger and Ingels 2018; Zeppilli et al. 2018).

And yet, mainstream benthic biological research still travels other routes, while meiobenthos is not really recognized by the general 'biology community' and its pertinent journals, instead often is considered irrelevant: the *'Meiobenthos Dilemma'* (Giere 2009, p. 422). Whether this dilemma is inherent or a result of today's alienation from nature remains arguable, but one reason for the frequent disregard of meiobenthos within benthos research is unawareness, unawareness of the considerable potential scientific work with small organisms can attain (see Schratzberger 2012).

© The Author(s), under exclusive license to Springer Nature Switzerland AG 2019
O. Giere, *Perspectives in Meiobenthology*, SpringerBriefs in Biology,
https://doi.org/10.1007/978-3-030-13966-7_1

A good example of this biased perception is the worldwide loss of species, the catastrophic destruction of biodiversity and, thus, of genetic information, by mankind. The essential discussion on this topic to which so many of us feel committed, is hardly ever documented by examples from the realm of meiobenthos. This is probably in part because of our limited level of meiofaunal species inventory, but certainly because disappearance of small, invisible parts of the ecosystems does not bother the broader public, is simply not realised. Even among meiobenthos research, this topic is underrepresented. You simply cannot reach people by the loss of nematode or harpacticoid species. This they have in common with many protists. What is their appeal compared to whales, fish or corals? Therefore, I decided to omit this aspect, when selecting topics that might have the potential to attract attention in the scientific or/and general public. Here I recommend publications on macrobenthos as more effective.

Understandably, not every field of meiobenthic research can have the potential for general scientific or societal relevance. Some research directions, although scientifically rigorous, will remain in the background. But to gain future relevance to both the science and the general public, the main thrust of our meiobenthic endeavours should move into fields exemplified here. In these benthological fields, the focus is frequently on macrofauna, especially when it comes to economically relevant species. Destroyed mussel beds or threatened coral reefs simply call for more attention than soft bottoms with their invisible meiofauna threatened by destruction or pollution. How then to attract attention? In order to better merge in the main stream, we should focus meiobenthos studies on aspects relevant for general biological and integrated environmental problems.

Many future directive trends have been suggested in the textbook on meiobenthology (Giere 2009). But ten years later, I wonder: Has the situation changed and has meiobenthology in the last decade left its corner and merged with the benthological mainstream? An overview of the recent literature shows many encouraging facets mostly originating from a few leading 'meiofauna schools'. Today considerably more contributions on meiofauna are published in top-ranked journals, most of them resulting from internationally cooperating teams. That, I feel, are positive signs. On the other hand and in accordance with Warwick's keynote speech at the 2016-International Meiofauna Conference (Warwick 2018), the situation is still unsatisfactory, with contributions being rather descriptive, too often asking 'what?' and 'where?' rather than 'why?' and 'how?' Only when we are embedding our factual knowledge on meiobenthos in cause analyses and when we put this in context with general theoretical, conceptual or societal aspects will we receive enough attention for a promising future. Realizing the present situation and considering our overall state of knowledge on meiofauna, I would advise to remain modest and in discussions perhaps avoid speaking of the 'meiofaunal biosphere' (Creer et al. 2010), the 'ecological quality status' of an area (Semprucci et al. 2018) or similar connotations promising far-reaching goals.

But can't one simply hide and remain in the tiny, carefree playground of meiobenthos research? Why all these efforts, why can't we stay longer where many of us are, doing their research properly, fascinated by the little pets in their tiny world? The disregard of meiofauna publications in the broader literature surveys expressed by Warwick (see above) shows the need for a change.

Any perceived lack of scientific and societal relevance will result in a decreasing support from financial sources with a shrinking employment rate of scientists as a consequence. A field that in many aspects is still young would then dwindle to irrelevance, although it houses so many unique animal groups with enigmatic structures and functions, a field of so many potentials!

In which realms can studies on meiobenthos become more significant for general benthology and even gain public awareness? This was the starting point of thoughts I pondered over many years. While the recent paper '*Meiofauna Matters*' by Schratzberger and Ingels (2018) mainly relates to ecology and ecosystem services and addresses also decision-makers, my present treatise has a broader, perhaps more didactic intention:

(a) Highlight those meiobenthological directions believed to be critically influential for the future and
(b) Illustrate these by summarizing and reflecting on some recent examples of such studies.

Any attempt of this kind is necessarily based on subjective opinion. But long-term experience with meiobenthos research, and acquaintance of leading researchers in the field, both old 'hats' and younger 'rising stars' alike, puts me in the fortunate position of being able to select appropriate topics and indicate some representative papers that could lead the way in these directions. This endeavour is favoured by the freedom as professor emeritus. Being no longer in any occupational 'rat race' and approaching my professional conclusion, I can observe the scenery from a more objective, qualifying distance. All of these aspects hopefully support well-balanced conclusions.

Before I specify those meiobenthic research fields that I consider of relevance in the sense outlined above, I like to explain my rationale for their selection on the one hand and the disregard of other studies on the other hand. Many careful papers of more local importance, often with most interesting topics, will have to be disregarded. This is, by no means, an indication of minor scientific value. But—using a comparison from harbour life—these works are the colourfully specialized meiobenthic convoy. They represent the flock of smaller boats that find their position in the wake of the strong tugboats. The tugboats are the ones to determine the general course and set the pace of progress while they lead their diverse followers with more local, but often precious cargo, behind them. The leading boats, being more generalists, avoid niches too narrow and will find pathways of broader attention. They are usually piloted by a few personalities that powerfully structure the nature, function and quantity of (scientific) trade.

Indeed, in many fields of meiobenthic research, the last decades have seen remarkable progress. With the broad adoption of new methods in disciplines as diverse as molecular genetics, ecology, nature protection, pollution, but also in classic taxonomy and evolution, meiobenthology has fundamentally changed. To me, this is particularly evident in ecological research, where automated sampling methods, molecular serial identification ('environmental' or 'meta-barcoding') and sophisticated computer-based evaluation methods are linking basic research with applied topics such as nature destruction, biotope preservation or diversity assessment—analyses in a complexity not imaginable some thirty years ago. Hence, the prerequisites are there

for a deeper understanding of functional processes, particularly of those connecting
the meio-world with its neighbouring micro- and macro-world.

Compared to the scientific presentations at the triennial conferences of the International Association of Meiobenthologists in the 1970s and 1980s, we are, so it
seems, indeed moving along a path already outlined by von Humboldt (1845):
the ever-changing 'entire scenery of nature' is only comprehensible by combining
detailed/local studies, connecting trends and comparing regions ('Alles ist Wechselwirkung' - everything is mutual interaction'). This demanding pathway towards
increasing complexity and coverage would not be possible without the electronic
interconnections of researchers worldwide. An example of this global cooperation is
the recent paper by Bluhm et al. (2018) on sea ice meiofauna that amalgamates data
from numerous researchers on a 'pan-Arctic' scale (Fig. 1).

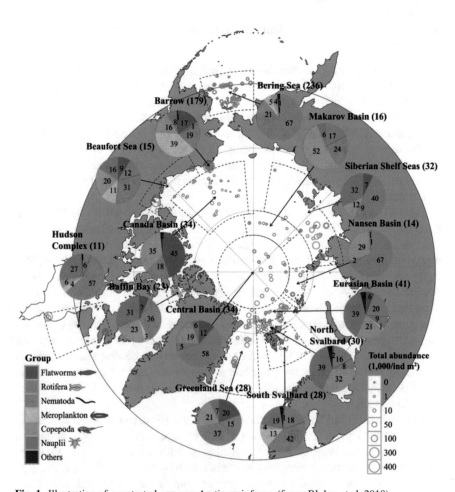

Fig. 1 Illustration of a metastudy on pan-Arctic meiofauna (from: Bluhm et al. 2018)

Comprehensive treatises and monographs often are particularly useful helpers serving the generalizing aims. They often level the road from meiobenthology to mainstream benthology, from species analysis to functional understanding of processes and further to societal services. There is another, important function inherent in general reviews: they combine textbook knowledge with results from fresh research, they are of high didactic relevance and they enable easier orientation in the complex field where tiny animals, often with unusual biological capacities, may attain important roles. Thus, they facilitate access to and attraction by meiofauna.

Because of this relevance, I would like to give three examples in this introductory chapter that well illustrate these characteristics.

- Rarely has the huge water body of the Mediterranean Sea been more comprehensively analysed under aspects of biodiversity than by Danovaro et al. (2010) *'Deep-Sea Biodiversity in the Mediterranean Sea: The Known, the Unknown, and the Unknowable'*. This paper highlights the close connection between meiobenthos and other faunal elements.
- The second directive paper comes from the field of applied sciences. The summarizing article of Montagna et al. (2013) *'Deep-Sea Benthic Footprint of the Deepwater Horizon Blowout'*, while using advanced evaluation methods, provides information that is comprehensible not only to specialists, but also to the public and the politicians. This paper covers the first large oil spill, which affected the deep-sea and left an unexpectedly grave benthic impact. It seems, this new type of oil spills requires a new type of study with a broad frame reaching beyond the borders of subject-specific assessments. Its profound analyses could serve as templates for future accidents to be expected.
- A central goal in general ecology, freshwater or marine, is revealing community patterns and their dependence on the environment: geographical variables, trophic relations and their structuring interactions. The article by Dümmer et al. (2016) *'Varying Patterns on Varying Scales'* evaluates data from numerous European lakes and connects them using mathematical correlations. It gives an exemplary overview over the impact of scale-related parameters on nematode distribution.

As different as the above topics may be, they and others in the chapters that follow show the breadth and diversity of meiobenthology. They can serve as examples of future, integrated research in meiobenthology, a scientific field that is both relevant and fascinating.

References

Bluhm BA, Hop H, Vihtakari M et al (2018) Sea ice meiofauna distribution on local to pan-Arctic scales. Ecol Evol 8:2350–2364

Creer S, Fonseca VG, Porazinska DL et al (2010) Ultrasequencing of the meiofaunal biosphere: practice, pitfalls and promises. Mol Ecol 19(Suppl):4–20

Danovaro R, Company JB, Corinaldesi C et al (2010) Deep-sea biodiversity in the Mediterranean Sea: the known, the unknown, and the unknowable. PLoS ONE 5(8): e11832, 25 pp

Dümmer B, Ristau K, Traunspurger W (2016) Varying patterns on varying scales: a metacommunity analysis of nematodes in European lakes. PLoS ONE 11(3):e0151866. https://doi.org/10.1371/journal.pone.0151866

Giere O (2009) Meiobenthology. The microscopic motile fauna of aquatic sediments, 2nd edn. Springer, Berlin, Heidelberg, 527 pp

Isimco (2016) Draft book of abstracts. In: 16th international Meiofauna conference, Heraklion, Greece, 3–8 July 2016, 182 pp

Montagna PA, Baguley JG, Cooksey C et al (2013) Deep-sea benthic footprint of the deepwater horizon blowout. PLoS ONE 8(8):e70540. https://doi.org/10.1371/journal.pone.0070540

Schmidt-Araya JM, Schmid PE, Tod SP, Esteban GF (2016) Trophic positioning of meiofauna revealed by stable isotopes and food web analyses. Ecology 97:3099–3109

Schratzberger M (2012) On the relevance of meiobenthic research for policy-makers. Mar Pollut Bull 64:2639–2644

Schratzberger M, Ingels J (2018) Meiofauna matters: the roles of meiofauna in benthic ecosystems. J Exp Mar Biol Ecol 502:12–25 http://doi.org/10.1016/j.jembe.2017.01.007

Semprucci F, Balsamo M, Appoloni L, Sandulli R (2018) Assessment of ecological quality status along the Apulian coasts (eastern Mediterranean Sea) based on meiobenthic and nematode assemblages. Mar Biodivers 48:105–115

von Humboldt A (1845) Kosmos. Entwurf einer physischen Weltbeschreibung. Die Andere Bibliothek. Berlin (2004), 935 pp

Warwick R (2018) The contrasting histories of marine and freshwater meiobenthic research—a result of differing life histories and adaptive strategies? J Exp Mar Biol Ecol 502:4–11

Zeppilli D, Leduc D, Fontanier C et al (2018) Characteristics of meiofauna in extreme marine ecosystems: a review. Mar Biodivers 48:35–71 doi.org/https://doi.org/10.1007/s12526-017-0815-z

Chapter 2
Fields of General Scientific Importance and Public Interest

1 New Areas, Novel Communities, Exotic Biotopes

Spectacular novelties in fascinating biotopes – The earth's crust – Anoxic depths – Ubiquitous biofilms – The world of groundwater, our health reservoirs

Each unexplored field, both in a geographical and in a figurative sense, attracts human curiosity and prompts new studies. Even in an electronically interconnected world with news and pictures circulating within seconds, so far unexplored faunas with novel appearances living in unknown, fascinating biotopes excite researchers and the broad public alike. Here, meiofauna seems to be advantaged: often occurring in great numbers, most of the recently discovered animals, even of high taxonomic ranks, have been of meiofaunal size.

Perhaps the most amazing meiofauna habitat worldwide, so far thought inaccessible and even uninhabitable to metazoan life, is the deep subsurface rock. In fractures some thousand metres below the surface, several species of nematodes were found thriving on rich bacterial biofilms in warm, hypoxic subterranean aquifers. Among them, a new *Halicephalobus* nematode with the inspirational iconic species name *'mephisto'* (Borgonie et al. 2011). This finding was a rare instance of meiofauna surfacing in the pages of most international scientific journals and even entering the public media. Nobody would have expected finding these *'frontiersmen of life'* in rocks deep in the earth's crust. The scientists even were able to do some experimental work on the trophic and ecophysiological attributes of these extremophiles. Not only did this study attract considerable public attention, it also expanded our conception of the metazoan biosphere. Does it also gain unexpected topicality after the recent discovery of a water lake deep in the Martian crust?

Another report of similar novelty and incredibility is the finding of meiobenthic organisms entirely isolated from the oxic world and completing their life cycle in total anoxia: a permanently anoxic, hypersaline deep-sea basin in the Mediterranean Sea is inhabited by specimens of new Loricifera species, which have been found in perfectly viable condition and in various developmental and reproductive stages

Fig. 1 Rich biofilms and bacterial mats at the floor around deep-sea hydrothermal vents, depths 3.017 m, Mid-Atlantic Ridge (Copyright MARUM, University of Bremen)

(Danovaro et al. 2010b). Although predicted for a long time, the authors could now provide 'compelling evidence' of the first (non-parasitic) metazoan existing under permanent anoxia and chemically hostile conditions. In addition, certain cytological features of these Loriciferans have been shown to cross a borderline between metazoans and protozoans: their cells lack mitochondria, but contain hydrogenosomes, mitochondrial derivatives usually found only in some protists and fungi from anoxic environments (see Sect. 1 in Chap. 5).

Life and evolution under isolation for millions of years—another spectacular attraction for studies of meiobenthic life: rich nematode and copepod communities have been found in a suboxic and sulphidic cave in Romania with a strange aquatic biocoenosis in complete darkness (Muschiol et al. 2015). What is the basis of this unusual persistence? Microbial biofilms (Fig. 1).

Biofilms, the '*complex communities of microbiota that live in association with aquatic interfaces*' (Weitere et al. 2018), are good examples of a divergent perception: studied since long by microbiologists, these ubiquitous biotopes seem as yet exotic to and little studied by meiobenthologists. Their ecological importance for meiofauna as grazing grounds often remains unrecognized. Cooperations between microbiologists and meiobenthologists combining chemical and microbiological technologies are about to reveal the central role of biofilms bringing them into the focus of ben-

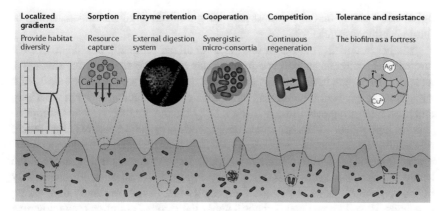

Fig. 2 Functional diversity of biofilms and their effects—a diagrammatic compilation (from: Flemming et al. 2016)

thic science. As turntable and mediator between prokaryotic and meiobenthic life, biofilms develop wherever rich bacteria, algae and protist populations are thriving (Fig. 1). Considered as the 'most successful forms of life on earth' (Flemming and Wingender 2010), biofilms serve as a habitat former, sediment stabilizer, a physical refuge, a highly productive food source and a hotspot of mutually interactive and regulative influences between bacteria, protozoans and metazoans (Fig. 2).

These 'fortresses of life' are the ubiquitous matrix of fundamental relevance for meiofaunal life (Van Colen et al. 2014; Flemming et al. 2016; Fig. 2). Devouring, agglutinating and structuring biofilms, meiofauna can attain regulative functions optimizing them as feeding grounds for bacteria. Secretions of marine and freshwater meiobenthos stimulate growth of microorganisms (bacteria, protozoans, fungi and microalgae), effects known since long and described as 'gardening' or 'engineering'. Cooperations between microbiologists and meiobenthologists, which combine new chemical and microbiological technologies, are about to underline the central role of biofilms for benthological studies (Majdi et al. 2012; Mialet et al. 2013). Pigment examinations in biofilms have confirmed the positive correlation between algae flourishing in these mucoid strata and abundance of meiofauna supporting the trophic role of mucus secretions. Meiofaunal muddy galleries and tubes compacted by biofilms and mucus secretions also favour colonization of macrobenthos.

All these aspects emphasize the central role of mucus secretions, both in marine and limnic habitats, as the structuring link between various benthic domains. Microbial and meiofaunal biofilms have even been ascribed a prominent role in the central biogeochemical processes with a high socio-economic value, such as benthos recruitment rates, water flow dynamics, sediment stability, even ecosystem engineering (Van Colen et al. 2014). Weitere et al. (2018) underline the role of biofilms connecting planktonic and benthic food webs. Continuation and intensification of the interdisciplinary, though complex, cooperation between microbiologists and meiobenthologists will become one of the most promising fields in a holistic understanding of future benthic research.

Another fascinating biotope of relevance for all our future life lies directly below our feet: the subterranean world of groundwater. Numerous aquifers, tiny streamlets and powerful rivers, hard to access domes and deep caves, are the hidden world of a surprising, specialized subterranean meiofauna with adaptations testifying to a long and most interesting evolutionary history. Stygofauna with their structural simplifications, high rate of endemism and heterogeneous distribution patterns often reflect past geological and climatic processes, spawning further studies of their general relevance (Galassi et al. 2009; Stoch et al. 2009). Beyond these scientific aspects, the health of unpolluted groundwater is of direct relevance for all of us. The sensitivity of the stygobiotic fauna to pollution, especially that of the frequently dominating copepods, is well known and perhaps a better indicator of pollution than numerous chemical tests. Considering the relevance of these biotopes as sources of our drinking water and the indicative potential of their meiofauna, far more studies are required. Since intensified studies will disclose many more groundwater species (Deharveng et al. 2009), meiofauna from groundwater layers have promising potentials as health indicators and offer a broad research basis on freshwater meiobenthology. Again, meiobenthologists are in the focus here, and their warnings will receive general attention regarding the frightening relevance of our potentially endangered drinking water resources (see Sect. 4 in Chap. 3).

2 The Deep-Sea—More Than an Interesting 'No Man's Land'

Little explored but soon to be exploited – Diversity, similarity and variability patterns as general indicators – Impact of surface conditions on deep-sea meiobenthos – Deep-sea in the focus of industrial exploitation – Meiofauna in a clue signal position

The depths of the oceans have always fascinated scientists and the public alike. And yet, in those depths many processes and recent scientific discoveries are far from understood. Hence, the deep-sea combines the attraction of unexplored faunas with an immense importance for surface life, for the climate and even for our planetary welfare. At the deep-sea bottom, seamounts, hadal troughs, hot vents and mud volcanoes are new research fields, and together, they represent the most vulnerable and least resilient biotopes on the globe. Their fauna often are compared with 'alien' beings, many of them undescribed. However, the deep-sea is not only a haven for taxonomists and ecologists. With locally rich deposits of valuable metals and rare earths, it has long been in the focus of the international mining industry—despite unresolved perils to the local environment and its fauna as well as to distant areas being impacted by huge turbidity clouds (for details and pictures of the massive machinery planned to use, see Internet 'SOLWARA 1'-Project 2008; 'Midas'-Project 2018).

In view of their high diversity and abundance, meiofauna can provide statistically valid data from relatively few and small samples. Useful both for answering general biological questions and indicating human impact, investigations of deep-sea meiobenthos can reveal principles of fauna distribution, diversity patterns and colonization processes. Thus, they are well suited both for answering general biological questions and disputed environmental cases.

It is a widely accepted paradigm that population density and alpha-diversity in the sea are conversely related to depth: decreasing food supply leads to decreasing faunal abundance and biodiversity with depth (in the Mediterranean Sea, the gradient of this decrease seems often particularly expressed). The harsh nutritive conditions in greater depths seem to favour a community of specialists, while ubiquitous 'generalists' become less abundant.

The copepod family Dirivultidae typically populates disjunct deep-sea hydrothermal vents all over the world. While, at the family level, their distribution is exceptionally wide, the genera and species are highly endemic (Gollner et al. 2010). This contrast prompted Plum et al. (2017) to perform experiments on copepods in vent fields in the Mid-Atlantic Ridge: moving away from hot vents, towards more normal deep-sea conditions, species number and dominance of non-vent specialists generally increased. Conversely, with increasing vent activities (toxic vent fluids, rising temperatures), dirivultids dominated regardless of the type of substratum.

The phenomenon of high meiofauna similarity despite separation by large distances has also been found in nematodes at various polymetallic nodule sites (Pape et al. 2017). This would confirm the contention that local deep-sea faunal diversity reflects similarity of prevailing conditions and would contradict the common tenet of a high and locally restricted diversity in the deep ocean.

In large-scale studies in the Mediterranean Sea, Danovaro et al. (2010a, 2013) searched for factors and patterns structuring the diversity of deep-sea fauna. Analysing numerous data sets from meiofauna (Foraminifera and Nematoda) and macrofauna, the conclusion was that sediment heterogeneity, differences in food supply, nutrient utilization as well as varying predation pressure seem to explain the diversity patterns found.

Biologically more meaningful than assessing species number per se are analyses of the complex relations between numerical diversity to functional diversity. For the Mediterranean deep-sea, Baldrighi and Manini (2015) showed that interacting competition and predation, which may change across size classes and taxa, apparently play the decisive role. Additionally, in this complex functional interplay, the choice of spatial scale is an important factor influencing diversity and confounds comparisons (Danovaro et al. 2013). Can these relations become generalized or are they of only local relevance? Is the deep-sea a highly diverse environment only at a local scale, while regionally this diversity becomes suppressed?

Large-scale distribution patterns of meiobenthos on the deep-sea bottom are, among other factors, influenced by hydrographical connectivity and community turnover. The underlying driving processes are important for understanding faunal recovery and, therefore, for assessing human interference. In a large- and multiple-scale study in the Atlantic deep-sea, Lins et al. (2017) concluded that environmental conditions in the depth are strongly influenced by processes at the surface and affect species turnover and community composition ('surface productivity regulates diversity'): a rich and stable food input seems to infer a lower alpha-diversity (species number) combined with a low turnover within stations, and a highly variable or low supply with resources would promote higher alpha-diversity. Hydrodynamic instability, i.e. a strong exchange across different depths or a connection of deep-sea water with that of shallower sites, promotes rapid exchange between different communities and enhances beta-diversity. These interchanges between different depths would also explain the scarcity of 'depth-endemic lineages' in the deep-sea. Rosli et al. (2018) also underline, in their profound review on the ecology of deep-sea meiofauna, the role of food supply and local oceanographic characteristics, at least for the small-scale variability. In order to generalize the results, we need, despite the considerable increase in deep-sea studies during the last decade, more of those large-scale evaluations for other oceans also. Another promising aspect for deep-sea studies is experimental work, which becomes increasingly applicable even in this remote world (see Freese et al. 2012; Zeppilli et al. 2015).

As much as deep-sea biology awaits further discovery and the ecology is little understood, one principle is clear: the deep-sea and its fauna is not an isolated and pristine world, undisturbed by human activities, as it was thought for a long time. Regarding the threats of these vast und vulnerable habitats by increasing accessibility to and interest in human exploitation, the scarcity of information on deep-sea fauna and ecology is particularly problematical.

This is, of course, closely linked to the recent progress in constructing sophisticated underwater vehicles of an unprecedented, electronic-based accuracy and a rapidly increasing working range. Specifically, the mining of valuable ores from ancient and recent hydrothermal vent sites and harvesting of polymetallic nodules are in the focus of investors from many nations (see review articles by Gollner et al. 2017; Jones et al. 2017; for pictures of machinery see Internet 'Nautilus Mining').

Although both substrata, nodule fields and vent bottoms, are of completely different origin and location, and although the diversity and natural variance of their fauna are little known, the results extracted from both of these comprehensive and detailed meta-analyses seem to confirm general deep-sea textbook trends when dealing with biological reactions to industrial disturbances. Impact severity depends on scale, duration (repetition), location (isolation, connectivity to the undisturbed surroundings) and fauna composition (small-sized and mobile animals less impacted than sessile macrofauna). And of course, any impact is exacerbated by the much slower turnover rates in the deep-sea compared to shallow waters and the generally high sensitivity to disturbance.

We have to learn much more about the underlying driving processes. Here, a wide and relevant field is open for further meiobenthological studies—meiofauna as a tool to better assess the impact of human interference and understand the range of faunal recovery.

Compiling these results and those from other papers, obvious generalizations are: after a massive impact, the general decline both in abundance and diversity may be followed by a shorter-term recovery (or even overshoot reactions) that may appear in some taxa with a higher turnover, larger resilience and/or effective distributional capacity. That means the comparison between meio- and macrobenthos seems to favour the smaller fauna in impact/disturbance studies. Although the results may differ with area, type of disturbance and depth, at present, one generalizable statement emerges from the above compilations: for deep-sea mining, we will have to consider general recovery times of at least several decades. Further details cannot be given yet in view of significant knowledge gaps in the studies: no information from fauna communities around inactive vents and many structural limitations in past studies.

My general conclusion: our insufficient information cannot yet provide a reliable basis for counteracting the pressure exerted by international economy and industry. This pressure is increasing since many contractors tend to equate the approaching end of the international exploration phase of polymetallic nodules with the start of possible exploitation activities. Under these circumstances, if biologically based and environmentally effective management protocols for forthcoming mining activities can be developed, the tools to use are studies on meiofauna. They can offer statistically relevant data important for strong juridical expertise. Thus, meiobenthic studies could gain further importance and perception even beyond their intrinsic biological interest. Requirements: effective evaluation, methods such as semi-automated fluid-imaging technologies or environmental barcoding (see Sect. 1 in Chap. 6). In addition, each polymetallic nodule and hydrothermal vent site, populated by macro- and meiobenthic animals, offers a good opportunity for cooperation of meio- and macrobenthologists in documenting these unique biomes with their fragile resilience and uncertain mode of recolonization.

Concluding, all benthological studies in the deep-sea, even those on meiobenthos, are facing two caveats.

(1) Results have to be mostly based on a few replicates only. Regarding the high variation in species abundance and diversity within just the area of one multi-corer, any extrapolation to large-scale or long-term patterns and any conclusions drawn from those figures should be regarded with scepticism (Gallucci et al. 2009).
(2) Studies focussed on single taxa may not be representative of whole assemblages and generalization might be misleading: pioneering species may recover and even exceed pre-disturbance levels after a short time, while the community composition as a whole may remain affected and diversity reduced for decades.

3 Newly Accessible Polar Regions

Polar regions, open for meiobenthic research – Soon affected by human use? – Diversity of meiofauna as usual? – Relevance of local studies – First polar pollution studies urgent

Polar regions and the deep-sea have aspects in common: both are becoming increasingly accessible in recent years, both seem biologically sensitive having a biota of low metabolic rate and little resilience and both are of high public and commercial interest. With the continuing retreat of the Arctic and Antarctic ice shields, more and more areas of wide, untouched sediment, most of them unexplored, become open to research—but probably earlier to human contact. This situation is strikingly mirrored by a recent press release of E. Hunke (Los Alamos National Laboratory, USA): 'The polar regions are not desolate, they're actually alive with shipping, energy development, fishing, hunting, research and military defense operations' (Marine Link, Dec. 5, 2018).

Macrobenthic animals in these 'new' polar regions often occur in low densities. This is in contrast to many meiofauna. A range of different studies indicate that all major meiofaunal groups known from temperate shallow sediments are well represented in the polar zones. Nematodes and annelids, but also rotifers, harpacticoids and foraminiferans are important ecosystem members (Brannock et al. 2018; Fonseca et al. 2017; Hardy et al. 2016; Zeppilli et al. 2018). Nevertheless, on lower taxonomic levels, the specificity of polar meiofauna seems remarkable, comprising many undescribed species. In barcode units from Antarctic samples, Fonseca et al. (2017) found only a small percentage (1–4%) of meiofauna fully matching the previously known data sets. Comparison with other barcoding studies of the Antarctic meiofauna showed considerable variations in the diversity of taxa. The higher diversity recordings of polar meiofauna reported by Brannock et al. (2018) may have methodological reasons, such as the limited level of identification still prevailing in electronic databanks. Considering the insufficient resolution at lower taxonomic levels, it seems too early to conclude that diversity changes might also reflect differences in sediment composition, depth and season.

The differing meiofauna composition in polar ecosystems aside, what are the functional relations among macro- and meiobenthos in these regions? To what extent are meiofauna important for nutrient regeneration and what are the interactions with microfauna? The early study by Urban-Malinga and Moens (2006) underlined the high rate of organic mineralization (detritus production), especially in beaches with coarse sand. However, the meiobenthic share in this process has been calculated to be rather modest (<5% consumption by meiofauna), while microbes seemed to play the major role. The links of meiofauna to macrobenthos in polar regions have apparently not yet studied.

In parallel to other 'empty biotopes' that become accessible to faunal colonization, Pasotti et al. (2015) describe from Antarctic sediments a relation between time span of glacier retreat and the functional diversity (complexity of trophic interactions) among the meiobenthos. The sediments that only recently became ice-free (some 15 years ago) hosted an impoverished community of species that occupied a 'wide isotopic niche' with a low trophic redundancy, typical r-strategists, dominated by colonizer species. In contrast, in communities populating older ice-free sites, the trophic niches became narrower and their trophic spectrum more interconnected. This would indicate a more intensive biomass transfer and resource recycling. The authors conclude from this difference a good capacity of polar (meio-)faunal communities to adapt to the rapidly ongoing climatic changes in their environment.

The sympagic interstitial fauna in the network of brine channels in the ice will also undergo changes with respect to the global temperature rise (Kiko et al. 2017; Bluhm et al. 2018). While old and thick ice shields often harbour endemic taxa with poor swimming ability and dispersal capacity (e.g. most nematodes), the spreading thinner ice plates that form seasonally are preferred by populations of 'colonizers' that have resting stages, good swimming ability and a high general tolerance (ciliates, rotifers, turbellarians, only few nematodes). Hence, it seems that the classical ecological principles of pioneering and conservative species, of ubiquitists and specialists, maintain their validity also under polar conditions. Rose et al. (2015) even identified some corresponding traits between 'new' polar and deep-sea meiofauna. However, these similarities seem to be local, and to result from parallel ecological conditions such as food limitation and retarded biological processing. Other studies confirmed marked meiofaunal differences between shallow and deep-sea sites in the Arctic, a feature also typical for other regions (Fonseca et al. 2017).

In general, life conditions and meiofauna in the newly accessible polar regions may be as heterogeneous, locally and seasonally varying as those in temperate regions. For instance, sympagic meiofauna in seasonally formed ice are more variable in composition than those from permanent ice shields. Differing results underline the need for future detailed investigations with a consistent methodology and coordinated logistics (Bluhm et al. 2018). Which timescales, which grades of resilience and recovery are to be expected under the extreme polar conditions? Do the local processes described in some few pioneering studies, allow already for generalization on the vast polar continents? As in the deep-sea, the enormous extent of polar regions can only be reasonably studied for community composition and ecosystem functioning by methods allowing high sample throughput, e.g. by modern molecular genetic approaches (see Sect. 1 in Chap. 6). If well-standardized these new methods will enable a holistic examination of polar meiofauna and eventually allow for extrapolations on general trends in the communities.

Signs of pollution in polar seas, primarily by petroleum hydrocarbons and heavy metals, are increasing with increasing traffic and industrial activities. One of the first experimental studies in polar regions on the effects of different petroleum hydrocarbons (Stark et al. 2017), exemplary in its long-term and differentiated approach (five years, four fuel mixtures), showed nematodes and harpacticoids to have different responses to and recovery rates from oil pollution. Even after five years, the abundance and alpha-diversity remained reduced, while functional parameters, such as feeding type composition and maturity index, showed first trends of recovery. This study not only underlines the long-lasting noxious effects of such pollutants under polar climatic conditions, but the diverging types and temporal scales of reactions are a serious warning: when assessing effects of pollution we cannot rely too much on short-term, single-species tests. More meiofauna surveys as well as experimental work in these areas are imperative for reliable data serving governmental authorities, the shipping industry and tourism.

In parallel to the deep-sea, meiofauna research in these new polar habitats is not only another rewarding, widely 'empty' scientific field, but it also urgently calls for investigation. Why urgently? So far, in many parts of the polar caps, pristine conditions prevail, but the recent global warming increasingly allows industries and commercial shipping access enhancing the concomitant risk of pollution. Considering the political and societal pressure on these areas, the time frame available for adequate research reactions, in this case meiofauna studies, is extremely narrow.

References

Baldrighi E, Manini E (2015) Deep-sea meiofauna and macrofauna diversity and functional diversity: are they related? Mar Biodiv 45(3):469–488. https://doi.org/10.1007/s12526-015-0333-9

Bluhm BA, Hop H, Vihtakari M et al (2018) Sea ice meiofauna distribution on local to pan-Arctic scales. Ecol Evol 8:2350–2364

Borgonie G, García-Moyano A, Litthauer D et al (2011) Nematoda from the terrestrial deep subsurface of South Africa. Nature 474(7349):79–82. https://doi.org/10.1038/nature09974. PMID 21637257

Brannock PM, Learman DR, Mahon AR et al (2018) Meiobenthic community composition and biodiversity along a 5500 km transect of Western Antarctica: a metabarcoding analysis. Mar Ecol Prog Ser 603:47–60. https://doi.org/10.3354/meps12717

Danovaro R, Dell'Anno A, Pusceddu A et al (2010a) The first metazoa living in permanently anoxic conditions. BMC Biol 8:30–40

Danovaro R, Company JB, Corinaldesi C et al (2010b) Deep-sea biodiversity in the Mediterranean Sea: the known, the unknown, and the unknowable. PLoS ONE 5(8):e11832, 25 pp

Danovaro R, Carugati L, Corinaldesi C et al (2013) Multiple spatial scale analyses provide new clues on patterns and drivers of deep-sea nematode diversity. Deep Sea Res II Topical Stud Oceanogr 92:97–106

Deharveng L, Stoch F, Gibert J et al (2009) Groundwater biodiversity in Europe. Freshwat Biol 54:709–726

Flemming H-C, Wingender J (2010) The biofilm matrix. Nat Rev Microbiol 8:623–633

Flemming H-C, Wingender J, Szewzyk U et al (2016) Biofilms: an emergent form of bacterial life. Nat Rev Microbiol 14:563–575

Fonseca VG, Sinniger F, Gaspar JM et al (2017) Revealing higher than expected meiofaunal diversity in Antarctic sediments: a metabarcoding approach. Sci Rep 7:6094. https://doi.org/10.1038/s41598-017-06687-x

Freese D, Schewe I, Kanzog C et al (2012) Recolonisation of new habitats by meiobenthic organisms in the deep Arctic Ocean: an experimental approach. Polar Biol 35:1801–1813

Galassi DMP, Huys R, Reid JW (2009) Diversity, ecology and evolution of groundwater copepods. Aquat Sci 54:691–708

Galluci F, Moens T, Fonseca G (2009) Small-scale spatial patterns of meiobenthos in the Arctic deep sea. Mar Biodivers 39:9–25

Gollner S, Ivanenko VN, Arbizu PM, Bright M (2010) Advances in taxonomy, ecology, and biogeography of Dirivultidae (Copepoda) associated with chemosynthetic environments in the Deep Sea. PLoS ONE 5(8):e9801. https://doi.org/10.1371/journal.pone.0009801

Gollner S, Kaiser S, Menzel L et al (2017) Resilience of benthic deep-sea fauna to mining activities. Mar Environ Res 129:76–101. http://dx.doi.org/10.1016/

Hardy SM, Bik HM, Blanchard AL (2016) Assessing benthic meiofaunal community structure in the Alaskan Arctic: a high-throughput DNA sequencing approach. North Pacific Research Board (NPRB) Project 1303 Final Rep 39 pp

Jones DOB, Kaiser S, Sweetman AK et al (2017) Biological responses to disturbance from simulated deep-sea polymetallic nodule mining. PLoS ONE 12(2):e0171750. https://doi.org/10.1371/journal.pone

Kiko R, Kern S, Kramer M, Mütze H (2017) Colonization of newly forming Arctic sea ice by meiofauna: a case study for the future Arctic? Polar Biol 40:1277–1288

Lins L, Leliaert F, Riehl T et al (2017) Evaluating environmental drivers of spatial variability in free-living nematode assemblages along the Portuguese margin. Biogeosciences 14:651–669

Majdi N, Tackx M, Buffan-Dubau E (2012) Trophic positioning and microphytobenthic carbon uptake of biofilm-dwelling meiofauna in a temperate river. Freshwat Biol 57:1180–1190. https://doi.org/10.1111/j.1365-2427.2012.02784.x

Mialet B, Majdi N, Tackx M et al (2013) Selective feeding of bdelloid rotifers in river biofilms. PLoS ONE 8(9):e75352. https://doi.org/10.1371/journal.pone.0075352

MIDAS' Summary EC-report Nr 603418 (2018) Seascape Consultants Ltd., UK, 28 pp

Muschiol D, Giere O, Traunspurger W (2015) Population dynamics of a cavernicolous nematode community in a chemoautotrophic groundwater system. Limnol Oceanogr 60:127–135

Pape E, Bezerra TN, Hauquier F et al (2017) Limited spatial and temporal variability in meiofauna and nematode communities at distant but environmentally similar sites in an Area of Interest for deep-sea mining. Front Mar Sci 4:206. https://doi.org/10.3389/fmars.2017.00205

Pasotti F, Saravia LA, De Troch M et al (2015) Benthic trophic interactions in an Antarctic shallow water ecosystem affected by recent glacier retreat. PLoS ONE 10(11):e0141742. https://doi.org/10.1371/journal.pone

Plum C, Pradillon F, Fujiwara Y, Sarrazin J (2017) Copepod colonization of organic and inorganic substrata at a deep-sea hydrothermal vent site on the Mid-Atlantic Ridge. Deep Sea Res II 137:335–348

Rose A, Ingels J, Raes M et al (2015) Long-term iceshelf-covered meiobenthic communities of the Antarctic continental shelf resemble those of the deep sea. Mar Biodivers 45:743–762

Rosli N, Leduc D, Rowden AA, Probert PK (2018) Review of recent trends in ecological studies of deep-sea meiofauna, with focus on patterns and processes at small to regional spatial scales. Mar Biodivers 48:13–34

SOLWARA 1 Project (2008) Main report Vol A and B, environmental impact statement. Nautilus Minerals, NL

Stark JS, Mohammad M, McMinn A, Ingels J (2017) The effects of hydrocarbons on meiofauna in marine sediments in Antarctica. J Exp Mar Biol Ecol 496:56–73

Stoch F, Artheau M, Brancelj A et al (2009) Biodiversity indicators in European ground waters: towards a predictive model of stygobiotic species richness. Freshwat Biol 54:745–755

Urban-Malinga B, Moens T (2006) Fate of the organic matter in Arctic intertidal sediments: is utilisation by meiofauna important? J Sea Res 56:239–248

Van Colen C, Underwood GJC, Serôdio J, Paterson DM (2014) Ecology of intertidal microbial biofilms: mechanisms, patterns and future research needs. J Sea Res 92:2–5

Weitere M, Erken M, Majdi N et al (2018) The food web perspective on aquatic biofilms. Ecol Monogr. https://doi.org/10.1002/ecm.1315

Zeppilli D, Vanreusel A, Pradillon F et al (2015) Rapid colonisation by nematodes on organic and inorganic substrata deployed at the deep-sea lucky strike hydrothermal vent field (Mid-Atlantic Ridge). Mar Biodivers 45:489–504

Zeppilli D, Leduc D, Fontanier C et al (2018) Characteristics of meiofauna in extreme marine ecosystems: a review. Mar Biodivers 48:35–71. https://doi.org/10.1007/s12526-017-0815-z

Chapter 3
Pollution and Meiofauna—Old Topics, New Hazards

New types of pollution – Analyses with new goals and new methods – Advantages of working with meiofauna – Problems of 'bulk identification'

Regrettably, pollution, in one or another form, will remain a permanent hazard even in the future. So, what is new with studies on pollution, where is the forward perspective? In the last years, new developments caused new sources of pollution with a risk potential of an unprecedented, global scale: water acidification through climate change and environmental impact of plastic debris. Other pollutants envisaged here are indeed 'old', but their possible fields of impact are new and require novel studies in hitherto untouched reaches: petroleum hydrocarbons widespread at the deep-sea floor and in groundwater layers.

Compared to earlier single-factor analyses, modern impact studies are much more demanding. They should consider ecosystem services and even qualify as the basis for environmental surveys and administrative decisions. In serving these high goals, they often aim to assess the 'ecological status' of an entire region, even to 'develop an ecosystem model'. As yet, this seems to me a grand, but still illusionary ambition. In any case, a more holistic view of pollution impacts will support the relevance of studies on meiofauna communities. However, it is a much more challenging task compared to the traditional single-species tests. Hence, modern pollution analyses often comprise a broad frame of community parameters. This integrative goal provides a more realistic and generalizable evaluation: a complex faunal community seems to react more sensitively and mirrors field conditions better than laboratory tests (Hägerbäumer et al. 2016). There is a compromise between assessing the complex reactions of ecosystems and the rather artificial 'single-species settings': microcosm experiments. Laboratories using meiofauna for microcosm approaches take advantage of the 'high abundance and diversity scenario' that can be established with meiobenthos at small spatial scales (Santos et al. 2018). These provide, beyond data on critical toxicity levels, information about differences at the species level and behavioural reactions of concomitant species. But one has to take into account that evaluation of multi-species experiments is a highly demanding goal.

O. Giere, *Perspectives in Meiobenthology*, SpringerBriefs in Biology, https://doi.org/10.1007/978-3-030-13966-7_3

Pollution studies based on meiobenthos have often been challenged by one obstacle: tedious identification work that requires the expertise of specialists. Breaking down the rich number of meiofauna specimens to a low taxonomic level (genera, species if possible) is a prerequisite for meaningful interpretation of pollution-induced changes in species richness, diversity or toxicity in general. An example : As nematodes and harpacticoid copepods will regularly be encountered, even in the severely polluted sites, only a detailed taxonomic differentiation can allow for assessing well-definable changes in faunal composition, e.g. survival of the more hardy ubiquitists and disappearance of the sensitive specialists. Only differentiation will allow determining taxa of indicative value. It may represent an old-fashioned conception of biology and pollution ecology, but sorting meiofauna to the level of orders or higher ranks ('bulk identification') does not have the same informational value than basing on genera or species. With unspecific data, even refined computer-assisted statistical methods will not differentiate between dramatic losses in sensitive species, while other, opportunistic species of the same numerical range (or higher) with a high reproductive potential will thrive. When a better-resolved meiofauna identification (morphological or molecular) is not feasible, careful and detailed chemical sediment analyses for toxic substances might render more conclusive results.

Recent developments in molecular tools are valuable steps towards solving the previous problem of meiofauna identification. Methods such as DNA taxonomy, environmental barcoding, meta-barcoding, high-throughput sequencing, clustering and fingerprinting (see reviews by Brannock and Halanych 2015; Fonseca et al. 2017; Fontaneto et al. 2015) yield a plethora of data that can be combined with the potential of electronic databanks (e.g. NEMYS, WORLD FORAMINIFERA DATABASE). With their broad pool of information, these new tools enable studies on an integrative scale with the potential of indicating both general ecological trends and reactions to pollution. Hence, provided that some pitfalls are considered (see Creer et al. 2010), the impact of pollutants on meiofauna composition can be assessed today in wide investigation areas, at reasonable timescales and by manageable teams at relatively low costs (see review by Zeppilli et al. 2015).

Those impact studies looking especially at effects on connected, mostly interactive biological processes such as growth, reproduction, population dynamics and competition, benefit from the privileged and inherent potential of meiobenthos compared to macrobenthos: numerous species per area with high individual numbers, rapid reactivity, short generation times, easy sampling at reasonable costs. Accordingly, marine nematodes have recently been suggested as candidates for the standardized 'Ecological Quality Assessment' (Semprucci et al. 2016). Determination of indicator species using promising mathematical procedures is another approach for selection of suitable methods (Stoch et al. 2009; Turner and Montagna 2016).

Out of the multitude of polluting scenarios, inadequate to cover in this essay, a few will be pointed out here, which are considered particularly threatening to our future welfare. Although often well noticed and covered by the media, they remain little understood and will probably harass us during the forthcoming decades despite all advanced knowledge and sophisticated methods. As diverse as they are, their investigation has one feature in common: pertinent studies often appear biased by societal, economic or political arguments and conditions. This does not facilitate an independent, objective assessment.

– Water acidification by CO_2 increase
– Microplastics and plastic fibres
– Oil spills at the deep-sea floor
– Petroleum hydrocarbons and groundwater aquifers.

The fact that the following remarks mostly relate to the marine realm may be ascribed to the proclivity and prevailing experience of the author. By no means must it be interpreted as an indication that the world of freshwater is and will become less threatened by human activities.

1 Water Acidification by CO_2 Increase

More complex effects than just the declining pH – Calcareous meiofauna suitable test objects – Experiments with intrinsic problems – Accompanying chemical and physiological reactions important

Among the effects of climate change that gain a perilous relevance not only for mankind are the global CO_2 increase and the concomitant water acidification. An expected ocean acidification of -0.3 to -0.4 pH units by the year 2050 sounds rather abstract and unimpressive to the public, but expressed as % change from the baseline level in the 1980s, it means an increase in the hydrogen ion concentration of some 30%, while the acidity of the ocean will more than double in the next 40 years (see Wikipedia). Spontaneously and primarily, one would envisage the noxious impact on carbonate dissolution, but beyond that, rising CO_2 levels cause changes in the complex carbonate chemistry of the oceans, a field as yet little understood. The biological reactions of benthic communities to increased CO_2 concentrations (hypercapnia) probably become exacerbated by the concomitant global warming and enhanced problems of oxygen deficiency. In combination, the alteration of these overarching environmental factors may elicit ecosystem changes of an unprecedented globality, with alarming consequences for the diversity of life, sometimes compared with mass extinctions in earlier geological periods (Bellard et al. 2012).

Seawater acidification, so it was contended, would change life primarily in the water column. But there is no reason to assume that processes in the overlying water do not also influence life in the aquatic sediments. Here, the impact becomes strongly modified by the multiple chemical processes, by the metabolic cycles of microbes, by mucus layers and biofilms typical for the intricate microtexture of the interstitial space (see Sect. 1 in Chap. 2). Recent studies show that acidification does affect the benthic system as much as the open waters, specifically when it comes to the buffering effects of calcareous sediments and their bottom fauna. Besides economically important shelled molluscs and crustaceans, the spectacular coral reefs with their environmental relevance are in the focus. But the structural complexity of these colonial organisms and their indirect development, including non-calcareous larval stages, aggravates detailed studies. Yet, as evidenced in the review article by Byrne (2011), studies are still focussing on macrobenthos. And they rarely consider the interactive effects of thermotolerance and calcification potential under elevated acidity.

Here, calcareous meiofauna such as the Foraminifera and Ostracoda, ubiquitous in coralline sands, offer easier objects for analyses on a micro-scale—another reason for using meiofauna as preferred study objects. Under structural, distributional and population-ecological aspects, their rich and speciose stocks could serve as valuable indicators of acidification effects. Comparisons of hard-shelled with soft-shelled foraminiferan species would further allow for conclusions on specific carbonate-dissolution processes. In addition, the well-documented fossil record of both meio-faunal groups would provide data from previous time periods in comparison with contemporary findings and render the studies a new time dimension. This brings studies on (sub)fossil meiofauna into our focus. Barker and Elderfield (2002) studied foraminiferan tests from glacial/interglacial periods and demonstrated that lowering of ambient carbonate concentrations was paralleled by a decrease in shell weight.

Experimental set-ups face the problem of correctly mirroring the situation in the field where subtle changes in pH and PCO_2 with longer-lasting endurance are characteristic. Experiments usually encompass short time periods of some weeks only (e.g. Rassmann et al. 2018). Therefore, in order to prove for effects, one often applies more drastic incremental steps (e.g. Meadows et al 2015: from 8.0 stepwise down to 6.7). Are the recorded manipulative reactions under these conditions directly caused by acidification or by the harsh experimental conditions themselves, e.g. by changes in the microbial community composition, perhaps also by trophic deterioration? For these reasons, various experiments with oftentimes drastic changes in pH (e.g. Ricketts et al. 2009—nearly 4.0 units) seem of limited value to mirror the impact of CO_2 sequestration in sediments. Can they indicate valid responses, especially when complex ecological parameters (richness, diversity, dominance) are analysed?

Experiments based on microstructural and physiological details often yield a more realistic answer. Applying PCO_2 levels that correspond to those likely occurring in future decades to a benthic foraminiferan species, McIntyre-Wressnig et al (2013) demonstrate the complexity of symptoms that would indicate damage: While growth and survival were not directly affected, the specimens regularly suffered from disso-lution of their tests. If this effect becomes generalized, it would point to considerable biological consequences.

What about Ostracoda under increased PCO_2 levels? Their calcification processes would give valuable inferences, especially in comparison with those in foraminifer-ans. For unclear reasons, pertinent studies on this speciose taxon, commonly inhab-iting coral reef sands, are rare, but they are needed.

The reactions to acidification reported from non-calcareous meiofauna are divergent (see Zeppilli et al. 2015; Sarmento et al. 2015, 2017; Ingels et al. 2018): experiments with nematodes displayed strongly negative responses (decreasing diversity and population size). Other authors found particularly harpacticoids and their nauplii to respond negatively (Oh et al. 2017) or the data remained ambiguously

(Rassmann et al. 2018). As in many impact studies, the faunal response to one single factor is hard to assess and correctly interpret. It depends very much on the experimental setup, on other water conditions, the nature of sediment used, the microbial biota establishing in the system (see below), the impact of light, time, etc. As usual, coastal meiofauna, adapted to more fluctuating conditions, will be less reactive to pH decrease than those from deep-sea areas. Corresponding experiments by Kurihara et al. (2007) seem to confirm this.

Flume experiments with elevated CO_2 and/or temperatures in a realistic range (Ingels et al. 2018) showed less impact on meiofauna than expected. Since negative responses varied with sediment type, one might speculate about the impact of factors apparently not reported but influenced by sediment structure such as decrease in oxygen and pH concentrations. Also, in another microcosm experiment (North Sea intertidal flat meiofauna, short-term, slightly lowered pH and elevated temperature), the results are not coherent (Mevenkamp et al. 2018): copepod numbers reduced, nematodes stable, some rare groups affected. Hence, both experimental assessment and, more so, prediction of impacts by seawater acidification/temperature elevation on community composition must be considered with caution. We are far from understanding the impact of the multiple fluctuations and factorial interactions occurring under natural conditions. We have not yet arrived at generalizable conclusions about meiofauna sensitivity to CO_2.

In this situation, studies on the long-term reaction of meiofauna exposed to natural CO_2 vents and shallow-water vents might serve as an advantageous proxy for ocean acidification experiments (Dahms et al. 2018; Molari et al. 2018). Easily accessible shallow-water vents can serve as natural laboratories studying impacts of ocean acidification in particular and of global climate change (increased hydrographic extremes) in general. The comprehensive study by Molari et al. (2018) showed that at pH levels lowered by 0.3–0.4 units from natural (which might correspond to a predicted future ocean acidification), the nematode community showed a considerable functional change: reduced diversity, trend towards smaller, more r-selected species. These results became endorsed by transplant experiments exposing sands with their meiofauna from 'normal' locations to those of naturally lowered pH conditions near CO_2 seepages. One can easily imagine that these results can be generalized and could perhaps better mirror future real-life changes under acidified conditions than short-term experiments (Fig. 1).

In sediments, chemical fluxes of acidity and alkalinity can strongly deviate from those in the overlying water, since chemical sediment conditions are strongly mediated by the metabolism of microbial communities (Braeckman et al. 2014). In incubation experiments (14 d) of coastal sandy sediments, a decrease of 0.3 pH caused drastic fluctuations in the community oxygen consumption and benthic nitrification rates. On the other hand, the sediment column often exerts a buffering effect: in long-term experiments with natural sand, pore water calcium ion concentration remained close to saturation even when the overlying water was highly Calcium-unsaturated (Haynert et al. 2014).

The influence of the chemical community parameters mentioned above demonstrates that water acidification does not only affect calcification processes. Increased

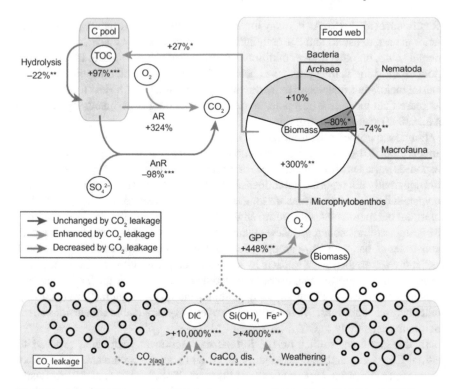

Fig. 1 Impact of long-term natural CO_2 leakage—a compilation (from: Molari et al. 2018)

acidity of the sediment system may also cause chemical changes in the specification of waterborne metals. Free-ion concentration of metals in sediments can have severe effects (Pascal et al. 2010). These side aspects of acidification might be relevant especially for the directly exposed meiofauna. These implications, usually not realized by biologists, urgently need intensified studies and consolidated results.

In nature, acidification is never a single driver, it is always linked to other factors, e.g. the relation to increasing global warming. Experiments should therefore incorporate at least increase in temperature plus acidification as a factorial pair. Moreover, indirect effects, often based on changed species interactions such as food competition, predator-prey relations or feeding rates, seem to act concurrently. In macrobenthos research, this has been demonstrated in several recent publications and impressively highlighted by experiments (King and Sebens 2018). In a recent comprehensive literature evaluation, Clements and Darrow (2018) found feeding activity of arthropods little affected 'in an acidifying ocean' compared to the more susceptible molluscs and echinoderms, and, of course, juveniles were more sensitive than adults. As yet, corresponding studies specifically on meiofauna do not yield conclusive results that unequivocally exclude other impact sources than acidification (Rassmann et al. 2018). Because this topic is so critically observed by the public, pertinent studies would have high relevance and draw much attention and comment, even beyond the specialists' niche. But again, in all of these publications, ecological reactions are described, but in-depth physiological pathways are rarely touched (see Chap. 5).

In his seminal overview related to the physiology of macrobenthos, Pörtner (2008) points out another aspect to be observed when dealing with ocean acidification. Numerous physiological pathways (e.g. protein synthesis, ion transport) depend on defined cellular pH levels and are not to be changed without major disorder and detrimental consequences. The predicted increasing variability in basic factors such as temperature, salinity and pH can elicit strong fluctuations in physiological and ecological processes to which small-bodied invertebrates due to their high tissue diffusion rate are particularly reactive. What is it that refrains specialists from starting with meiofauna? Meiofauna in their high reactivity, numerical abundance and diversity could give valuable answers to physiological effects of seawater acidification and concomitant factors.

2 Microplastics and Plastic Fibres

Ubiquitous microplastics – Ecological and metabolic risks – Uptake processes hardly studied – Impact on meiofauna? – Studies of high future relevance

Recently, the Ellen MacArthur Foundation predicted that by the middle of this century, the plastic load drifting in the oceans will weigh more than all fish populations combined (Fig. 2). This perilous and conspicuous source of pollution causes growing concern worldwide, and consequently, financial support for adequate studies remediating this threat is increasing.

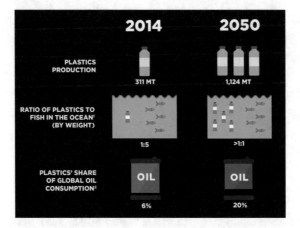

Fig. 2 Plastic versus fish in future oceans (from: Report Ellen MacArthur Foundation for WORLD ECONOMIC FORUM, Annual Meeting Davos 2016, modified)

While the public is horrified by moribund birds and turtles entangled in plastic nets, while tourism suffers wherever seashores become littered with plastic debris, many scientists are more alarmed by the mostly invisible and omnipresent load of plastic microparticles (below 5 mm in size). As 'primary microplastics', these represent nanobeads manufactured as additives to many cosmetic or household products. The 'secondary microplastics' originate from degrading, ground or abraded larger plastic wastes, from colour films and rubber layers, ropes and nets, also from antifouling compounds. Particularly, microfibers from synthetic textiles (non-woven fabric) are surrounding us and have been shown to occur in all oceans and depths including the deep-sea floor (Van Cauwenberghe et al. 2015b; GESAMP-Report 2015, 2016; Taylor et al. 2016; Barrows et al. 2018; Paul-Pont et al. 2018; Fig. 3). Especially in Arctic areas, often believed pristine and exempted from human impact, both the deep-sea sediment (Bergmann et al. 2017) and the ice surface carry an unimagined

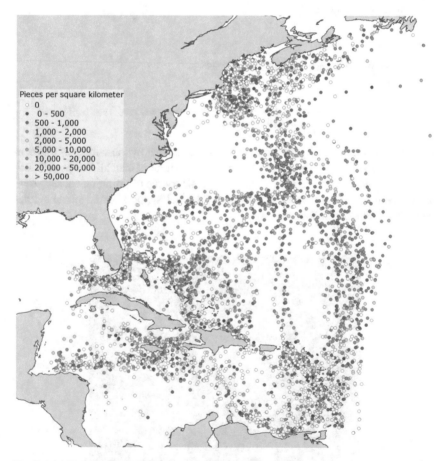

Fig. 3 Distribution and density of microplastics in Western Atlantic waters (from: GESAMP 2015, with permission of IOM, modified)

load of microplastic particles. Sea ice in particular seems an unexpected sink of anthropogenic debris, which becomes distributed all over the globe by drifting and melting processes (Peeken et al. 2018).

The general risks of plastic wastes in the marine environment are summarized in some recent reviews (Li et al. 2016; Wang et al. 2016). But the studies addressing microplastics in sediments focus on macrofauna of commercial value or of 'engineer function' in the benthic ecosystem, definitely not on meiofauna (Van Cauwenberghe et al. 2015a; Paul-Pont et al. 2018). Therefore, although trying to describe the perils of microplastics in the realm of meiobenthos, I will have to primarily mention studies on macrobenthos with only some few comments on meiobenthos.

The ecological effects of microplastic particles on aquatic fauna are as diverse as the complex processes of their physicochemical and microbiological modification and degradation (see review by Paul-Pont et al. 2018). Biofouling, fragmentation and sorption play an important role. These impacts are often indirect and become recordable only at longer term (Fig. 4). When ingested at lower concentrations, deposit-feeding aquatic organisms mostly excrete the plastic particles by defecation without noxious (short-term) effects since intestinal layers are usually well adapted to removal of microscopic particles of roundish shape.

However, if microplastics occur in mass aggregations, their uptake by benthic and planktonic animals has been found to reduce the nutritional value resulting in weight loss, deteriorate the energy budget and cause an imbalanced behavioural activity (Wegner et al. 2012; Wright et al. 2013; Cole et al. 2015). Their indirect and

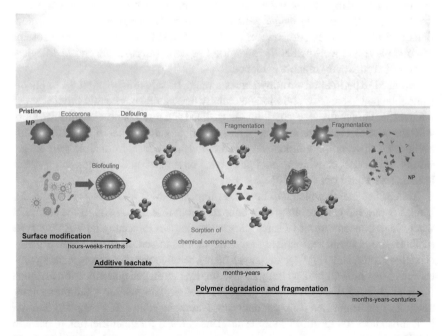

Fig. 4 Physicochemical and biological fate of microplastic particles in aquatic environments (from: Paul-Pont et al. 2018)

longer-lasting subsequent metabolic consequences may be even more serious (Van Cauwenberghe et al. 2015a; Paul-Pont et al. 2016): chemical softeners and noxious contaminants adsorbed to the particles' surface might become concentrated after ingestion (Andrady 2011). Bio-accumulated, they may generate ecotoxicological effects (reduced fecundity and other metabolic alterations, Wright et al. 2013; Green et al. 2016; Welden and Cowie 2016). Transferred along the food chain, the noxious impact on other aquatic species, fish and fish-eaters will escalate.

On the other hand, in a summarizing critical evaluation based on a broad survey of the literature, Koelmans et al. (2016) concluded that microplastic ingestion by marine organisms per se does not seem to have additional deleterious effects compared to those exerted by the load of organic chemicals to which they already are exposed in their environment. Also the recent study of Beiras and Tato (2019) on filter-feeding microplankton is questioning the widely assumed role of the common polyethylene microplastics as vectors of noxious organic contaminants in the sea.

However, all conclusions on a limited harm of microplastics may be taken cautiously regarding the inadequate number of histological and (longer-term) physiological studies. In the suspension-feeding blue mussel (*Mytilus edulis*), the uptake of microplastics elicited cellular lesions and pathological impacts in tissues and body fluids such as the blood and lysosomal systems (Moos et al. 2012). Visualized by cytochemical biomarkers, this study impressively showed the alarming and rapid tissue damage and other detrimental effects.

A dangerous threat, at least to macrobenthos, seems particularly caused by the ingestion of microfibers. Abraded from plasticware, cables and rods, these often needle-sharp microparticles today occur in every water body and are less easily excreted after uptake. They tend to penetrate tissue surfaces causing irritations, infections and other injuries (Barrows et al. 2018). Considering the economic role of seafood harvested worldwide, there are too few solid microscopic and histological studies of comparably reliable evidence.

We need experimental scrutiny across a range of scientific disciplines to understand this 'new' source of pollution potentially threatening our health: Studies under environmentally realistic, if possible standardized and reproducible conditions. Only when properties and concentrations of plastic particles are comparable, their bioavailability and distribution similar and the methods applied reproducible will be get results and models convincing enough to exert an impact on th public and on stakeholders (see Paul-Pont et al. 2018).

This brings meiobenthos into the focus: impact studies, especially at the cellular level, are particularly meaningful in animals living directly on and in the sediment where high concentrations of microplastics are known to accumulate. Appropriate methods are at hand: enrichment and extraction from sediment with high efficacy (Hidalgo-Ruz et al. 2012; Claessens et al. 2013), reliable and convenient staining procedures using bio-imaging techniques and, finally, automated quantification methods (Cole et al. 2015; Erni-Cassola et al. 2017). Rather than assessing the momentary amount of microplastics from gut analyses, Karakolis et al. (2018) applied fluorescent substances to digestible microbeads. This allowed for quantification of particle uptake with time.

Moreover, maintenance of great numbers of nematodes and harpacticoids in exper-imental jars and through several short generations would allow for more stringent methods and yield more significant results than comparable studies on some few mussels or lobsters. Provided that experimental exposure to microplastics stays in an environmentally realistic range (see warnings by Lenz et al. 2016), comprehen-sive impact bioassays with meiobenthos can be performed with relative ease and with empirically sound results. They would not only attract considerable attention, they will, according to recent reports of national science administrations, also attract appropriate financial support.

However, the present research situation on microplastics and meiofauna is weak and the data record poor. Among the few studies, the deposit-feeding polychaete *Sac-cocirrus* has been shown to intensively ingest microparticles such as fibres (Gusmão et al. 2016). But, in contrast to reports on macrobenthos (see above), this annelid seems to egest them without apparent physical harm. Yet, as pointed out above, we need exact and detailed micro-analyses on the cellular uptake of microplastic par-ticles. How and with which consequences are they transported through biological membranes? On the ecophysiological side, we need to assess the potential chemi-cal intoxication and the trophic imbalance caused by microplastics. Meiofauna and microplastics—an open research field in its infancy!

3   Oil Spills at the Deep-Sea Floor

Contamination of a pristine and sensitive ecosystem – New challenges with more deep-sea drilling worldwide

Oil pollution? An old hat with numerous meiofauna studies since 50 years! Even with the increasing use of environmentally neutral 'green' power, the future industrial production of numerous chemical substances will keep up the demand for petroleum hydrocarbons. Therefore, crude oil will remain a sought-after product for long times to come. However, the experience from many decades of maritime oil transportation led to advanced and mandatory technical precautions in the event of oil spills. Today, these can be more effectively avoided and confined, their fate and effects are fairly well covered and their damages better assessed. So, the damage rate has decreased considerably and mitigation actions are more effectively organized. This is valid for the classical oil tanker accident at the water surface.

But, worldwide oil drilling at the numerous offshore deep-water wells is a new challenge and pollution of deep-sea bottoms will predictably become a lasting prob-lem in many oceans. So, new physical, chemical and logistic problems arise. In contrast to surface conditions, in deep water, the lighter and particularly toxic hydro-carbons dissolve under pressure to form hydrates. The resulting complex mixture stays as a cloud suspended in the depth or may as a bottom layer cover the seafloor, probably with long-lasting, but as yet unexplored persistence and impact on the fauna (Fig. 5).

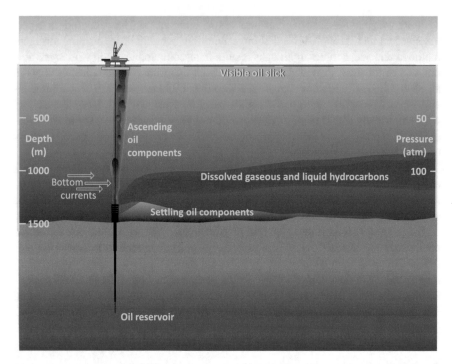

Fig. 5 Fate of petroleum hydrocarbons spilled under deep-sea conditions; distances not to scale (from: Oceanus Magazine WHOI, Oct. 2014, modified F. Widdel)

Under deep-sea conditions, the dominant pathway of oil degradation is bacterial biodegradation. But, whichever bacteria are interacting in the degradation chain (see Hu et al. 2017), studies on a successional pattern that plausibly would mirror the effectiveness and velocity of bacterial oil decomposition can only be realistic under the prevailing deep-sea conditions. However, the basic limitations are well known: under high pressure and low temperatures, bacterial activities are extremely low and slow.

The Deepwater Horizon blowout in 2010 was unique in many respects: it was the first very large spill occurring at the deep-sea bottom, it was spectacularly covered and put in scene by the media and, most importantly, it was, from the beginning, thoroughly pursued by microbiological and meiobenthological researchers using modern technical tools and evaluation methods (Bik et al. 2012; Montagna et al. 2013; Baguley et al. 2015; Hu et al. 2017). The results were not only published in high-profile scientific journals, they have also found access to and much attention by the public media.

Detailed and impressively, the data not only show severe impacts on the abyssal community structure with 'dramatic shifts', they also surprise by the size of the impacted zone (about 150–170 km^2). Even four years after the blow-up, losses of taxonomic richness and diversity in macro- and meiobenthos were significant while some signs of recovery were noticeable. Considering the effects on sediment bioturbation, organic matter decomposition and other functional aspects in the deep sea, these losses also mean impairment of whole ecosystem. Combined, the results of these studies indicate that full recovery of the impacted deep-sea ecosystem will be a matter of years to come (Montagna et al. 2013; Baguley et al. 2015; Reuscher et al. 2017).

Following up and explaining the intricate shifts in community structure, which the investigations revealed, would presumably need a longer-lasting and more refined assessment than any simple presence/absence calculation could indicate. Additionally, there is an increasing need to translate biophysical and ecological impacts into ecosystem services (fisheries, tourism, etc.) using mathematical models. 'Monetizing' the damages in nature in combination with corresponding legal initiatives may represent an effective way of reducing oil (and other) environmental perils.

Beyond the meiobenthic scope, the changes in taxonomic and functional interrelations have also relevant impact on the microbiota. Bik et al. (2012) showed that bacterial and fungal assemblages are probably decisive for the local ecosystem functioning. These subtle detrimental effects cannot be ascertained by standardized chemical analyses. Hence, there is an important chance and mission of meiofaunal and microbial studies to underline the indicative role of these organismic units.

Also, with respect to dispersant application, the Deepwater Horizon blowout was unprecedented: for the first time, these 'mitigating' chemical solvents were injected at the deep-sea bottom at some 1500 m depth (Reddy and Arey 2017). So far, results of the impact of this treatment on the meiofauna have not been found published—a serious gap considering the vivid debates about this treatment when oil is spilled at the surface.

After the disastrous consequences of the Deepwater Horizon accident and after the enormous attention by the media and public accompanying it, one should expect that every comparable accident would evoke highest alert and effective control measures by all drilling companies. However, under the title '*the biggest oil leak you've never heard of*', Eco Watch by Greenpeace (2016) informs about platform accidents with disastrous and ongoing oil spills in subsurface and deeper waters in the Gulf of Mexico. Without being stopped or adequately announced to the public, deep-sea oil pollution in enormous amounts seems to continue here. Considering the worldwide intensive offshore oil exploration in the deep ocean floor off Brazil, Uruguay and India, accidents comparable to Deepwater Horizon are immanent and seem inevitable. With the decreasing rate of oil spills in surface waters, our scientific efforts to evaluate the ecological consequences of oil pollution now have to be directed to the deep sea. This task is more urgent than ever, because, as long as oil is exploited, 'there will be spills' (EcoWatch by Greenpeace 2016).

4   Petroleum Hydrocarbons and Groundwater Aquifers

Fracking: More toxicants for more oil – Stygofauna as health indicators of drinking water

Hydraulic fracturing for oil and gas ('fracking'), the recent source of unexpected wealth, is in complete contrast to marine deep-water drilling: it affects a little noticed, even less explored and understood environment—the aquifers in the *terra firma*, deep beneath our feet, the source of our drinking water! While on the European continent fracking is widely banned, the present US government is a firm believer in its grand future, in the benign nature of methods used and in the integrity of the companies involved. So, mostly in North America, fracking will have a longer-lasting future and urgently needs to be tested by independent scientific groups.

Every insider familiar with the oil and gas production by fracking knows that even under regular conditions, traces of toxic chemicals are getting released, firstly by the chemicals added to the 'stimulation fluids' pumped down, then by the flowback of wastewater and the huge amount of flushing water that is required and eventually disposed. These sources have been shown to contaminate the surroundings. Although the fracking procedure itself is fairly local, this regular 'operational' contamination can affect wide areas when becoming distributed by the groundwater aquifers. The result would be benzene, aromatics, heavy metals, even glutaraldehyde and other substances in layers serving as groundwater strata.

Incriminating results on toxic contaminations of groundwater have been published early on (Warner et al. 2013; Kovats et al. 2014), but the scarcity of 'neutral', independent studies on the impact of fracking for oil and gas on the environment and the surrounding groundwater resources is a systemic and societal failure. The reports of the American Environmental Protection Agency (US EPA 2016) and by DiGiulio and Jackson (2016) might be considered a reliable basis giving a valuable overview, but their analyses focus on chemical and medical risks. While there exist studies on a potential chemical pollution of groundwater aquifers, there are only very few biology-based studies on the possible impact of fracking (e.g. Chen et al. 2017), and it seems that an unbiased (meio-)benthological study does not exist.

Hence, this is a big task and opportunity for future meiobenthos research, since the potentially threatened groundwater aquifers are the biotopes of stygobiotic, meiobenthic animals. The stygofauna, known for their sensitivity to pollutants, do not seem to be studied in detail in those regions, comparing geological strata exposed to fracking activities with unaffected ones. Methods suggested by Stoch et al. (2009) for the identification of indicator species in European groundwater could serve as a valuable guide for urgently needed unbiased meiofauna projects on the impact of fracking. This research remains an open field with an immense influence, especially in the USA where fracking is prospering.

References

Andrady L (2011) Microplastics in the marine environment. Mar Pollut Bull 62:1596–1605

Baguley JG, Montagna PA, Cooksey C et al (2015) Community response of deep-sea soft-sediment metazoan meiofauna to the Deepwater Horizon blowout and oil spill. Mar Ecol Prog Ser 528:127–140

Barker S, Elderfield H (2002) Foraminiferal calcification response to glacial–interglacial changes in atmospheric CO_2. Science 297:833–836

Barrows APW, Cathey SE, Petersen CW (2018) Marine environment microfiber contamination: global patterns and the diversity of microparticle origins. Environ Pollut 237:275–284

Beiras R, Tato T (2019) Microplastics do not increase toxicity of a hydrophobic organic chemical to marine plankton. Mar Pollut Bull 138:58–62

Bellard C, Bertelsmeier C, Leadley P et al (2012) Impacts of climate change on the future of biodiversity. Ecol Lett 15:365–377

Bergmann M, Wirzberger V, Krumpen T et al (2017) High quantities of microplastic in Arctic deep-sea sediments from the HAUSGARTEN observatory. Environ Sci Technol 51:11000–11010

Bik HM, Halanych KM, Sharma J, Thomas WK (2012) Dramatic shifts in benthic microbial eukaryote communities following the deepwater horizon oil spill. PLoS ONE 7(6):e38550. https://doi.org/10.1371/journal.pone.0038550

Braeckman U, Van Colen C, Guilini K et al (2014) Empirical evidence reveals seasonally dependent reduction in nitrification in coastal sediments subjected to near future ocean acidification. PLoS ONE 9(10):e1081 53. https://doi.org/10.1371/journal.pone.0108153

Brannock PM, Halanych KM (2015) Meiofaunal community analysis by high-throughput sequencing: comparison of extraction, quality filtering, and clustering methods. Mar Genom 23:67–75

Byrne M (2011) Impact of ocean warming and ocean acidification on marine invertebrate life history stages: vulnerabilities and potential for persistence in a changing ocean. Oceanog Mar Biol Ann Rev 49:1–42

Chen SS, Sun Y, Tsang DCW et al (2017) Potential impact of flowback water from hydraulic fracturing on agricultural soil quality: metal/metalloid bioaccessibility, microtox bioassay, and enzyme activities. Sci Total Environ 579:1419–1426

Claessens M, Van Cauwenberghe L, Vandegehuchte MB, Janssen CR (2013) New techniques for the detection of microplastics in sediments and field collected organisms. Mar Pollut Bull 70:227–233

Clements JC, Darrow ES (2018) Eating in an acidifying ocean: a quantitative review of elevated CO_2 effects on the feeding rates of calcifying marine invertebrates. Hydrobiologia 820:1–21

Cole M, Lindeque P, Fileman E et al (2015) The impact of polystyrene microplastics on feeding, function and fecundity in the marine copepod *Calanus helgolandicus*. Environ Sci Technol 49(2):1130–1137

Creer S, Fonseca VG, Porazinska DL et al (2010) Ultrasequencing of the meiofaunal biosphere: practice, pitfalls and promises. Mol Ecol 19(Suppl 1):4–20. https://doi.org/10.1111/j.1365-294X.2009.04473.x

Dahms H-U, Schizas NV, James RA et al (2018) Marine hydrothermal vents as templates for global change scenarios. Hydrobiologia 818:1–10. https://doi.org/10.1007/s10750-018-3598-8

DiGiulio DC, Jackson RB (2016) Impact to underground sources of drinking water and domestic wells from production well stimulation and completion practices in the Pavillion, Wyoming, field. Environ Sci Technol 50:4524–4536. https://doi.org/10.1021/acs.est.5b04970

EcoWatch by Greenpeace (2016) https://www.ecowatch.com/the-biggest-oil-leak-youve-never-heard-of

Erni-Cassola G, Gibson MI, Thompson RC et al (2017) Lost, but found with Nile Red: a novel method for detecting and quantifying small microplastics (1–20 μm) in environmental samples. Environ Sci Technol 51(23):13641–13648

Fonseca G, Fontaneto D, Di Domenico M (2017) Addressing biodiversity shortfalls in meiofauna. J Exp Mar Biol Ecol 502:26–38. http://dx.doi.org/10.1016/j.jembe.2017.05.007

Fontaneto D, Flot J-F, Tang CQ (2015) Guidelines for DNA taxonomy, with a focus on the meiofauna. Mar Biodiv 45:433–451

GESAMP (2015) Sources, fate and effects of microplastics in the marine environment: a global assessment (Kershaw PJ (ed)) (IMO/FAO/UNESCOIOC/UNIDO/WMO/IAEA/UN/UNEP/UNDP Joint group of experts on the scientific aspects of marine environmental protection). Rep Stud GESAMP No. 90, 96 p

GESAMP (2016) Sources, fate and effects of microplastics in the marine environment: part two of a global assessment (Kershaw PJ, Rochman CM, eds) (IMO/FAO/ UNESCO-IOC/UNIDO/WMO/IAEA/UN/UNEP/UNDP Joint group of experts on the scientific aspects of marine environmental protection). Rep Stud GESAMP No. 93, 220 p

Green DS, Boots B, Sigwart J et al (2016) Effects of conventional and biodegradable microplastics on a marine ecosystem engineer (*Arenicola marina*) and sediment nutrient cycling. Environ Pollut 208 B:426–434

Gusmão F, Di Domenico M, Amaral ACZ et al (2016) In situ ingestion of microfibres by meiofauna from sandy beaches. Environ Pollut 216:584–590

Hägerbäumer A, Höss S, Ristau K et al (2016) A comparative approach using ecotoxicological methods from single-species bioassays to model ecosystems. Environ Toxicol Chem 35:2987–2997

Haynert K, Schönfeld J, Schiebel R et al (2014) Response of benthic foraminifera to ocean acidification in their natural sediment environment: a long-term culturing experiment. Biogeosciences 11:1581–1597

Hidalgo-Ruz V, Gutow L, Thompson RC, Thiel M (2012) Microplastics in the marine environment: a review of the methods used for identification and quantification. Environ Sci Technol 46:3060–3075

Hu P, Dubinsky EA, Probst A et al (2017) Simulation of deepwater horizon oil plume reveals substrate specialization within a complex community of hydrocarbon degraders. PNAS 114(28). https://doi.org/10.1073/pnas.1703424114

Ingels J, dos Santos G, Hicks A (2018) Short-term CO_2 exposure and temperature rise effects on metazoan meiofauna and free-living nematodes in sandy and muddy sediments: results from a flume experiment. J Exp Mar Biol Ecol 502:211–226

Karakolis EG, Nguyen B, You JB et al (2018) Digestible fluorescent coatings for cumulative quantification of microplastic ingestion. Environ Sci Technol Lett 5:62–67

King W, Sebens KP (2018) Non-additive effects of air and water warming on an intertidal predator-prey interaction. Mar Biol 165:64. https://doi.org/10.1007/s00227-018-3320-4

Koelmans AA, Bakir A, Burton GA, Janssen CR (2016) Microplastic as a vector for chemicals in the aquatic environment. Critical review and model-supported re-interpretation of empirical studies. Environ Sci Technol 50(7):3315–3326

Kovats S, Depledge M, Haines A et al (2014) The health implications of fracking. Lancet 383(9919):757–758. https://doi.org/10.1016/S0140-6736(13)62700-2Cite

Kurihara H, Ishimatsu A, Shirayama Y (2007) Effects of elevated seawater CO_2 concentration on the meiofauna. J Mar Sci Technol 15:17–22

Lenz R, Enders K, Nielsen TG et al (2016) Microplastic exposure studies should be environmentally realistic. Proc Natl Acad Sci USA 113(29):E4121–E4122

Li WC, Tse HF, Fok L (2016) Plastic waste in the marine environment: a review of sources, occurrence and effects. Sci Total Environ 566–567:333–349

McIntyre-Wressnig A, Bernhard JM, McCorkle DC, Hallock P (2013) Non-lethal effects of ocean acidification on the symbiont-bearing benthic foraminifer *Amphistegina gibbosa*. Mar Ecol Prog Ser 472:45–60

Meadows AS, Ingels J, Widdicombe S et al (2015) Effects of elevated CO_2 and temperature on an intertidal meiobenthic community. J Exp Mar Biol Ecol 469:44–56

Mevenkamp L, Ong EZ, Van Colen C et al (2018) Combined, short-term exposure to reduced seawater pH and elevated temperature induces community shifts in an intertidal meiobenthic assemblage. Mar Environ Res 133:32–44

Molari M, Guilini K, Lott C et al (2018) CO_2 leakage alters biogeochemical and ecological functions of submarine sands. Sci Adv 4:eaao2040

Montagna PA, Baguley JG, Cooksey C et al (2013) Deep-sea benthic footprint of the deepwater horizon blowout. PLoS ONE 8(8):e70540. https://doi.org/10.1371/journal.pone.0070540

Moos Nv, Burkhardt-Holm P, Köhler A (2012) Uptake and effects of microplastics on cells and tissue of the blue mussel *Mytilus edulis* L. after an experimental exposure. Environ Sci Technol 46:11327–11335

Oh JH, Kim D, Kim TW et al (2017) Effect of increased pCO_2 in seawater on survival rate of different developmental stages of the harpacticoid copepod *Tigriopus japonicus*. Anim Cells Syst 21(3):217–222. https://doi.org/10.1080/19768354.2017.1326981

Pascal P-Y, Fleeger JW, Galvez F, Carman KR (2010) The toxicological interaction between ocean acidity and metals in coastal meiobenthic copepods. Mar Pollut Bull 60:2201–2208. https://doi.org/10.1016/j.marpolbul.2010.08.018

Paul-Pont I, Lacroix C, González Fernández C et al (2016) Exposure of marine mussels *Mytilus* spp. to polystyrene microplastics: toxicity and influence on fluoranthene bioaccumulation. Environ Pollut 216:724–737. http://dx.doi.org/10.1016/j.envpol.2016.06.039 10.1016/j.envpol.2016.06.039

Paul-Pont I, Tallec K, González-Fernández C (2018) Constraints and priorities for conducting experimental exposures of marine organisms to microplastics. Front Mar Sci 5:252. http://dx.doi.org/10.3389/fmars.2018.00252

Peeken I, Primke S, Beyer B et al (2018) Arctic sea ice is an important temporal sink and means of transport for microplastic. Nat Comm https://doi.org/10.1038/s41467-018-03825-5

Pörtner H-O (2008) Ecosystem effects of ocean acidification in times of ocean warming: a physiologist's view. Mar Ecol Prog Ser 373:203–217

Rassmann J, Lansard B, Gazeau F et al (2018) Impact of ocean acidification on the biogeochemistry and meiofaunal assemblage of carbonate-rich sediments: Results from core incubations (Bay of Villefranche, NW Mediterranean Sea). Mar Chem 203:102–119

Reddy CM, Arey JS (2017) Did dispersants help responders breathe easier? Chemical spray in Deepwater Horizon improved air quality at surface. Oceanus Mag WHOI 53(1), online Winter 2017

Reuscher MG, Baguley JG, Conrad-Forrest N et al (2017) Temporal patterns of deepwater horizon impacts on the benthic infauna of the northern Gulf of Mexico continental slope. PLoS ONE 12(6):e0179923. https://doi.org/10.1371/journal.pone.0179923

Ricketts ER, Kennett JP, Hill TM, Barry JP (2009) Effects of carbon dioxide sequestration on California margin deep-sea foraminiferal assemblages. Mar Micropaleontol 72:165–175

Santos ACC, Choueri RB, Pauly GDFE et al (2018) Is the microcosm approach using meiofauna community descriptors a suitable tool for ecotoxicological studies? Ecotoxicol Environ Saf 147:945–953

Sarmento VC, Souza TP, Esteves AM, Santos PJP (2015) Effects of seawater acidification on a coral reef meiofauna community. Coral Reefs 34:955–966

Sarmento VC, Pinheiro BR, Montes MJF, Santos PJP (2017) Impact of predicted climate change scenarios on a coral reef meiofauna community. ICES J Mar Sci. https://doi.org/10.1093/icesjms/fsw234

Semprucci F, Balsamo M, Sandulli R (2016) Assessment of the ecological quality (EcoQ) of the Venice lagoon using the structure and biodiversity of the meiofaunal assemblages. Ecol Ind 67:451–457

Stoch F, Artheau M, Brancelj A et al (2009) Biodiversity indicators in European ground waters: towards a predictive model of stygobiotic species richness. Freshw Biol 54:745–755

Taylor ML, Gwinnett C, Robinson LF, Woodall LC (2016) Plastic microfiber ingestion by deep-sea organisms. Sci Rep 6:33997

Turner E, Montagna PA (2016) The max bin regression method to identify maximum bioindicator responses to ecological drivers. Ecol Inform 36:118–125

U.S. EPA (2016) Hydraulic fracturing for oil and gas impacts from the hydraulic fracturing water cycle on drinking water resources in the United States (Final report). US Environmental Protection Agency, Washington, DC, EPA/600/R-16/236F

Van Cauwenberghe L, Claessens M, Vandegehuchte MB, Janssen CR (2015a) Microplastics are taken up by mussels (*Mytilus edulis*) and lugworms (*Arenicola marina*) living in natural habitats. Environ Pollut 199:10–17

Van Cauwenberghe L, Devriese L, Galgani F et al (2015b) Microplastics in sediments: a review of techniques, occurrence and effects. Mar Environ Res 111:5–17

Wang J, Tan Z, Peng J et al (2016) The behaviors of microplastics in the marine environment. Mar Environ Res 113:7–17

Warner NR, Christie CA, Jackson RB, Vengosh A (2013) Impacts of shale gas wastewater disposal on water quality in western Pennsylvania. Environ Sci Technol 47(20):11849–11857. https://doi.org/10.1021/es402165b

Wegner A, Besseling E, Foekema EM et al (2012) Effects of nanopolystyrene on the feeding behavior of the blue mussel (*Mytilus edulis* L.). Environ Toxicol Chem 31(11):2490–2497

Welden NAC, Cowie PR (2016) Long-term microplastic retention causes reduced body condition in the langoustine, *Nephrops norvegicus*. Environ Pollut 218 B:895–900

Wright SL, Rowe D, Thompson RC, Galloway TS (2013) Microplastic ingestion decreases energy reserves in marine worms. Curr Biol 23:R1031eR1033. http://dx.doi.org/10.1016/j.cub.2013.10.068

Zeppilli D, Sarrazin J, Leduc D et al (2015) Is the meiofauna a good indicator for climate change and anthropogenic impacts? Mar Biodiv 45:505–535

Chapter 4
Future Trend Lines in Ecological Meiobenthos Research

Ecology, key discipline in changing and threatened ecosystems – Meiofauna, the perfect bio-indicators – Hub position between microorganisms and macrobenthos – Interdisciplinary studies on ecosystem functions required – Standardisation and cooperation are prerequisites

In a world where human activities increasingly constrict, fragment and destroy natural habitats and threaten its animate nature, the study of ecosystem stability and resilience is becoming a dominant research discipline. Patterns of distribution, species inter-actions and modes of recolonization are underlying keys for our understanding of the dynamics in most environmental processes. The functioning of ecosystems, the complicated and ever-changing network of mutual organismic relations and the eco-logical quality status will be in the focus.

In the benthic realm, ecological research on meiofauna can attain a determining position by far exceeding its earlier role. Today's theoretical, statistical and technical instruments allow for more effective sampling, and computer-based logistics render an easier evaluation of meiofauna, which grant an increased predictive potential (see below). In this framework, custom-made and cost-effective molecular techniques play a central role. Combined with powerful microscopical analyses and digital calculations, the methodological progress alleviates the complex interpretation of data and relations (Fonseca et al. 2010, see Sect. 1 in Chap. 6).

This basis provides studies on meiobenthos a potential not envisaged in earlier times. Just because of their low motility and frequently rapid reactivity, these minute organisms, directly exposed to changes in their benthic environment, are prime bio-indicators. Thus, combined with corresponding data on abiotic parameters, microorganisms and macrobenthos, meiobenthic species can attain a 'hub' position. Due to their abundance and diversity, they 'can provide great benefit for metacom-munity analysis' (Dümmer et al. 2016).

No wonder that meiofauna work today mostly does not study and interpret single ecological relationships among isolated species in a locally restricted area. Instead, many relevant studies try to understand general patterns and processes, to identify ecosystem key functions and specify their regional correlates. Connected with neigh-bouring disciplines (cooperation with microbiology, mathematics) and based on a

O. Giere, *Perspectives in Meiobenthology*, SpringerBriefs in Biology,
https://doi.org/10.1007/978-3-030-13966-7_4

complex theoretical framework, interdisciplinary studies on meiofauna will increasingly address urgent problems of general relevance both for science and for common welfare:

a. Aspects of biodiversity as a basis for evaluating 'ecological quality'
b. Principles of faunal distribution, dispersal and resilience
c. Organismic interactions—meiofauna between microbiota and macrofauna.

Of course, the long-term objectives underlying these topics are challenging: understanding the 'functioning of (meiobenthic) ecosystems'. As yet, this often-cited goal is in most cases too complex to be reached. Perhaps it even betrays an unrealistic, mechanical view on living organisms? But it may stand as a euphemistic ambition directing our future efforts.

1   Aspects of Biodiversity

Functional diversity rather than biodiversity – The interrelations? – Acquiring diversity and connectivity patterns by use of mathematical models – large scale analyses with biogeographic aspects

In many studies on meiobenthos, it has become common to equate species number per habitat/area with 'diversity' or also 'biodiversity'. In ecological terms, this just signifies one aspect of diversity, 'alpha-diversity'. Often considered as an ecologically valuable index, does this invitingly simple concept really imply relevant community parameters? On the one hand, this biodiversity often relates to basic parameters such as sediment heterogeneity or food supply. Yet beyond that, biodiversity is influenced and determined by a framework of complex interactions: sustainability, resilience, niche width or 'system functioning' are relevant aspects. Hence, species numbers, often treated as a quality per se, gain their relevance only in context with environmental or ecological parameters. 'Natural history and experimental ecology must cooperate to help untangle ecological complexity, merging biodiversity with the functioning of ecological systems' (Boero and Bonsdorff 2007).

Why then became the calculation of 'biodiversity' in its simplified and limited meaning so popular? In contrast to assessment of species numbers by traditional scrutiny of diagnostic structures, species can now be determined by DNA taxonomy without even having to inspect a complete specimen. In their various approaches and extensions (meta-barcoding, environmental sequencing), these gene-based methods have gained so much practicability, cost-effectiveness and power (see Sect. 1 in Chap. 6) that in many studies they are taken as equivalent to a morphology-based determination of the 'biological value' of biotopes.

To a certain degree, this new assessment of biodiversity or species number can indeed serve as a basis for comparing faunas in different regions, depths or life conditions. Correctly interpreted, meta-barcoding can disclose distribution patterns fairly well, and these can even help to reveal community functions and their dynamics over time or environmental gradients. The changes in species number can indicate simple population parameters: negative or positive influences, impacts of food supply or pollutants. After evaluation of large-scale data (wider regions and higher numbers), indicator species can be derived, especially when using modern mathematical calculations (Stoch et al. 2009; Turner and Montagna 2016; Leasi et al. 2018).

However, even progressive barcoding methods, such as mass sequencing, are no magic tools. They suffer from theoretical and traditional handicaps. Theoretical: Where to draw the borderline between one genotypic sequence or taxonomic unit referred to as 'species' and its close neighbour? There remains an arbitrary margin known also from morphology-based taxonomy or the morphospecies concept.

The traditional (or better practical?) handicap: the vast number of 'operational taxonomic units' (OTUs) retrieved from mass sequencing or environmental sequencing becomes biased and restricted in its value for several reasons:

(1) The degree of alignment with classical and morphological voucher species is (as yet) limited (see Fonseca et al. 2010; Avó et al. 2017; Leasi et al. 2018). Hence, the high percentage of unassigned 'units' reported in many studies is certainly disappointing, but not surprising regarding the deficiency of taxonomically described species deposited in reference libraries. Even in well-studied areas, about one-third of all distinguishable genetic clusters were found to be undescribed (see Curini-Galletti et al. 2012; Costello 2015).

(2) Compared with morphological taxonomy, identification by meta-barcoding seems less accurate at low taxonomic ranks, while it works well in discriminating units at higher ranks (Leasi et al. 2018). Moreover, this comprehensive study disclosed that, depending on the method applied, the various meiofaunal taxa become differently resolved. In some groups, the recorded alpha-diversity was higher when identifying by traditional morphological taxa compared to the use of barcoding methods.

Oftentimes, studies on dominant meiofauna groups (e.g. nematodes), from marine and freshwater, from shallow and deep sites try to assess a potential relation between biodiversity (alpha-diversity) and functional diversity or even 'efficacy' of an ecosystem. It turns out that many of these complex concepts are scale dependent, i.e. factors relevant for smaller-scale investigations may change at greater scales (Dümmer et al. 2016; Rosli et al. 2017). Comparing alpha-diversity and functional diversity among nematodes at different depths in the Mediterranean, Baldrighi and Manini (2015) could not find a consistent relationship, but rather a varying dependence on size class within the taxon. Also in studies that compared meio- and macrobenthos, a parallel trend between alpha-diversity and functional diversity was rarely found (see Discussion in Baldrighi and Manini 2015). Since functional diversity, among other factors, strongly depends on the trophic spectrum, differences between meio- and macrofauna with their different composition of trophic types (e.g. scarce filter feeders in meiobenthos) are to be expected.

The frequently cited bathymetric gradient, typically found in sampling profiles from the surface towards the deep-sea, denotes an exponential decline in functional diversity compared to the more linear decline in numerical biodiversity (see Danovaro et al. 2008). While the dwindling food supply will be responsible for the general decrease in biodiversity, the exponentially diminishing functional diversity and the dropping of other 'ecosystem functions' are probably not to be explained by unifactorial influences and not valid for all diversity levels (Pusceddu et al. 2016).

The question is: Does a generally definable relation between numerical diversity and functional diversity exist? In nematodes, high species diversity did not relate to a greater functional diversity (Baldrighi and Manini 2015). Strong competition (niche overlap) as well as functional redundancy, typical for shallow sites, might reduce numerical biodiversity of assemblages but promote their functional diversity.

Altogether, many of the postulated relations, how ecological parameters influence biodiversity, have to be put into perspective (Leduc et al. 2013). Biodiversity depends on the spatial scale; it is strongly influenced by the maturity of the ecosystem, the prevailing limitations and by the abiotic situation. There may be loose ties only as well as direct interdependences. In essence, the entire ecological context, both abiotic and biotic, is of influence and hardly definable by a general relationship (Danovaro et al. 2013).

Scrutiny of data by complex mathematical models seems to be a promising approach when analysing these relations. Again, this requires cooperation of teams, sometimes disciplines, particularly when studying large areas. Analyses of biodiversity in large regions not only reveal distributional patterns and their influencing factors. They often allow for conclusions on the connectivity of communities and help to designate metacommunities. They thereby enable conclusions on biogeographical and evolutionary aspects. Altogether, determination of biodiversity by methods of 'integrative taxonomy', with molecular and modern morphological analyses combined, and by evaluation after comprehensive mathematical procedures (Fonseca et al. 2018) will become an increasingly powerful and reasonable tool. Basing on large species numbers, widespread meiofauna studies can assess quality of the environment, its resilience and its deterioration in a fairly objectified and verifiable procedure.

However, this optimistic conclusion is often constricted by the incoherent application of differing methods and scales with different underlying hypotheses. The resulting mosaic of diversity estimates does impede a reliable assessment of important ecosystem services such as predicting diversity changes under the impact of climate change. Described by Bellard et al. already in 2012, this scenario requests standardization of methods and cooperation among research groups, a demand as urgent as ever.

2 Reflections on Distribution, Dispersal and Colonization

Factors determining distribution, are they identical in meio- and macrofauna? – Driving factors for distribution? – The 'meiofauna paradox', is it real? – Distributional factors in marine and freshwater different? – Migration vs. dispersion – About ubiquitists, specialists and generalists and their fate in prospective conditions

Biodiversity is closely linked to distribution and dispersal. Thus, analyses of distributional principles are keystones for conservation biology, pollution threats, risk of invasive trends, etc. For these relevant assessments, can investigations using meiofauna with their numerous advantages (see previous chapter) provide significant progress? Does meiofauna and macrofauna distribution and diversity follow the same ground rules? One should think so, judging from many papers that draw conclusions on general distribution processes, regional or large scale while basing on meiofauna studies. However, do the distributional pathways of small benthic organisms with their limited means of active dispersion conform to those of macrobenthos? Comparing, for example, latitudinal/longitudinal trends in macro- and meiobenthos—can the connectivity of communities or the diversity relation to water depth (see above) indeed follow the same principles?

This problem also relates to the hitherto little understood 'meiofauna paradox' (see Giere 2009, p. 247), i.e. the distributional pattern of many meiobenthic populations, which, in the absence of diagnostically relevant morphological features, have been described as belonging to identical species despite their disconnected distribution sometimes extending even over separate continents. This ubiquity enigma of meiobenthos would contradict the existence of common distributional principles for benthos of all body sizes. In other words: Does this distributional model in meiofauna set limits to general characteristics common to both benthic communities? Under these circumstances, can meiofauna studies then really serve as a proxy of explanatory power also for distributional processes in macrobenthos?

Meiofauna and macrofauna experience their environment morphologically and physiologically in different ways and respond differently to habitat conditions, stress or pollution. Hence, both assemblages often react along different lines and may give different answers (Patricio et al. 2012). This in mind, any general interpretation of distribution patterns drawn from meiobenthic studies alone should be considered with appropriate care.

Direct comparisons of distribution patterns between macro- and meiofauna revealed that the responsible factors for distribution patterns are different and structure the benthos differently. Vanaverbeke et al. (2011) showed for the North Sea benthos that the meiobenthos (nematodes) distribution was mostly sediment related depending on factors influenced by sediment composition. In contrast, the macrofauna distribution was mostly related to water-column processes. In comparative freshwater studies on nematodes, it was the trophic state that, on a regional scale, could best explain distribution patterns. In contrast, on a larger scale, various geographical variables, e.g. watercourse patterns, were found responsible ('varying patterns on varying scales', Dümmer et al. 2016).

However, if we stay within the bounds of meiobenthos, analyses that outline distribution patterns and investigate their connectivity relations represent valuable tools for explaining biogeographical and ecological principles both in freshwater (lakes and river networks) (Dümmer et al. 2016) and in oceanic regions (Curini-Galletti et al. 2012; Lee and Riveros 2012). Many factors responsible for the distribution and spatial structure of meiofauna are since long literature-known, e.g. habitat heterogeneity (coarse sand vs. fine sand), food supply (shallow-water vs. deep-sea), hydrodynamics (lentic vs. exposed) or biological interactions (mobility, ontogenetic dispersal potential). The underlying task for the future will be to analyse the driving factors behind many results of a more descriptive character. Without understanding them, the core of many 'rules' remains unclear.

An example: The 'latitudinal range of biodiversity', one of the most interesting patterns, does it really relate to the underlying temperature gradient? As the increasing number of studies in polar latitudes show, a straightforward relation (alpha-diversity in high latitudes reduced, in warmer sites enhanced) does certainly not apply (see Sect. 3 in Chap. 2). A comparison of strictly parallel macro- and meiofauna samples from identical depths would give valuable insights into the causative factors and their general impact.

Which steps would be important for a deeper understanding? Certainly not just quoting biodiversity figures, but clarifying their relations, grasping the meaning of diversity, analysing connectivity patterns and faunal interactions—the drivers 'behind' are the important ones and would allow drawing parallels to macrobenthos and would contribute to a more general observance of meiobenthology. The most important tools to clarify these relevant variables, which often require very large data sets, are at hand now: competent use of high-standard molecular methods and refined statistical/mathematical evaluation.

The above-mentioned 'meiofauna paradox' can serve as a suitable example: The surprisingly wide distribution range of identical morphospecies (ubiquity hypothesis), does it hold up to scrutiny of genetic signatures? It becomes more and more revealed that in many meiofauna species, absence of morphological differentiation does not parallel genetic differentiation. Many of the famous meiobenthic cosmopolitan species are, in fact, complexes of cryptic species with distribution patterns reflecting a mosaic of genetic differences hidden under the mantle of a uniform morphology (Leasi et al. 2016). As genetic entities, many 'ubiquitous' taxa have much less uniformity and a clearly delimited distribution (Cerca et al. 2018; compare also Dümmer et al. 2016; Mendes et al. 2018). Also in Antarctic reaches, meta-barcoding analyses showed a marked genetic diversity in meiofauna from various regions along a long-distance transect (Brannock et al. 2018). This indicates a higher dispersal capacity of meiofauna than assumed. Hence, DNA-based analyses rather confirm the classical impact of geographical or ecological barriers for dispersion both in the marine and in freshwater environments. This would correspond to many macrofaunal patterns and connect both realms.

On the other hand, there remain intrinsic differences: compared to macrofauna, in meiobenthos active migration is negligible, but a meiofauna-sized organism gets easily passively dispersed by water currents and sediment drift, especially when living in the upper, more-exposed sediment layers (Boeckner et al. 2009; Cerca et al. 2018). Hence, in the open oceans, seamounts and isolated islands can serve as 'stepping stones' and their distributional role needs detailed scrutiny (Packmore and Riedl 2016). Further analysis of the somewhat contradicting factors relevant for the distribution of meiofauna (morphological uniformity, genetic isolation, ease of dispersion) requires more and perhaps experimental analyses regarding the principal modes of connectivity and dispersal in populations.

Scaling further down in organismic size, let us compare meiofauna with protists: In meiobenthic communities, Fonseca et al. (2014) concluded the distribution as resulting from a combination of dispersive influences and niche-driven processes, while in protists and microfauna, the 'ubiquity hypothesis' (see above) would mainly apply. The microbiota are the ones with the huge drifting potential, so they are rather ubiquitous and less isolated by distance.

Considering community characteristics, studies should also consider the 'maturity' of communities (sensu Bongers 1990). This paradigm tries to connect adaptive specialization and ecological niche width with the degree of speciation. Now the question arises: In a future world where rapid, often catastrophic environmental change and growing physiological stress is predicted to increase, which community is more fit for surviving the forthcoming threats—the more homogeneous one dominated by r-selected generalists or that of diversified, largely k-selected specialists? When spatial and temporal disturbances prevail, is a community with a high functional diversity and large number of k-specialists really more 'mature', can we assign a higher 'ecological quality' to it (Semprucci et al. 2016)? Aren't the ubiquitous generalists the kings of the future world? And anyway, perhaps, we should avoid these rather anthropocentric terms 'quality, maturity', etc.

3 Organismic Interactions—Meiofauna Between Microbiota and Macrofauna

Central role of food relations – Feeding types and their assessment – Role of diagnostic PCR – Fatty acid and carbon isotope identification relevant tools – Bacteria, sediment and meiofauna, a dynamic and interdependent complex – Symbioses, the closest interaction of bacteria and meiofauna

To comprehend a biocoenosis with its habitat, its organisms and multiple interactions requires more than assessing species numbers and patterns of distribution. The much more complex 'functional diversity' leads to a deeper understanding of how and where organisms are interacting. Without this demanding claim, we cannot properly realize the organisms' requirements and judge on the processes involved and the 'value' of a biotope.

The most evident interaction of an organism with its living environment, grazer, predator or prey alike, is its role in the food web. Most relations of meiobenthos both to the microbiota and macrobiota alike are comprised here. More than most other ecological aspects, they open the gate to a functional interpretation of biocoenoses. The feeding strategies and food relations are keys to many subsequent processes.

Understanding the tropho-ecological complexity is best achieved by direct observation and experimental work. However, in meiobenthology, only few papers (e.g. Hohberg and Traunspurger 2009; Mialet et al. 2013; Muschiol et al. 2015; Gansfort et al. 2018) are pursuing these often demanding and time-consuming tasks. Rather, conclusions on the nature and structure of relevant trophic pathways are based on the designation of feeding types, and these are usually ascribed to anatomical details of the mouthparts, a method that has been proved valuable in earlier decades (see Wieser 1953, for nematodes). However, can this indirect method really mirror the complexity of trophic interactions, especially their flexibility and resource dependence? Just these parameters are incompletely understood, poorly quantified and sometimes considered a black box (Moens et al. 2016).

In an ecosystem, the interactive connections, especially those with the microbiome, are a network of unimagined and variable ramifications, which put the above-mentioned concept of rather static, morphology-based feeding guilds into perspective. Instead, future food analyses use sophisticated sequencing methods, and they analyse fatty acid patterns (see Sect. 3 in Chap. 5) and natural stable-isotope relations. They will show an amazing trophic variability at the species level and, context-dependent, even among individuals.

But isn't the application of PCR-based methods in many meiofauna groups limited by preparatory difficulties due to the minute bodies and the minimal mass of their gut contents? Maghsoud et al. (2014), analysing the food spectrum of meiofaunal flatworms, solved the problem by their 'diagnostic PCR' method based on 18 s rDNA. In their seminal work, they could develop prey-taxon-specific primers that allowed to directly identify the prey of several 'turbellarian' species. Extension of this approach with a larger set of primers will reveal an amazing trophic width and flexibility also among other meiofaunal taxa. Applying these refined methods could explain patterns of direct coexistence of morphospecies and could question the conception of static feeding guilds based on identical mouth structures (Derycke et al. 2016). Thus, diagnostic PCR methods will not only diversify or even correct our conceptions of benthic food relations, they also have the potential for revealing entirely new meiofaunal food webs. This novel conception with many singular and mutual trophic interactions among meiofauna taxa calls for new approaches when categorizing trophic groups. They would lead to valid generalizations and allow comparisons with related species and, ultimately, with entire biocoenoses.

Another concept for determining the trophic position of meiofaunal taxa in ecosystems is the analysis of variations in the delta^{13}C or delta^{15}N signatures. Do meiofauna prefer grazing on autochthonous food (bacteria, biofilms, algae) or do they mainly live on allochthonous prey? Especially in freshwater habitats, the seasonal shift from autochthonous food (biofilms with lower delta values) in the summer to a prevalence of allochthonous prey (higher delta values) in fall and winter reflects the considerable dynamics in trophic relations. Measuring these parameters in the benthos of a lowland stream, Schmidt-Araya et al. (2016) found meiofauna to attain a surprisingly high position in the limnic food web. But in a dynamic system, can this position be maintained over the course of the year and be generalized? Maidi and Traunspurger (2017) showed that a 'general trophic switch' of carbon signatures is possible: although the meiofauna of a stream occupied various trophic levels, there was a seasonal trend towards the low-positioned compartment of fine particular organic matter.

Another approach for assessing the food spectrum of meiofauna is examining the meiofaunal fatty acid composition, often in combination with measurement of isotopic carbon signatures. In many cases, these biomarker methods, which connect trophic ecology and physiology (see Sect. 3 in Chap. 5), underline a wide and variable trophic range, starting either from microalgae (dinoflagellates, diatoms) or bacteria. Focussing on shallow-water harpacticoids, Leduc et al. (2009) found in experiments with meiofauna a high food uptake of polyunsaturated fatty acids that indicated the known trophic prevalence of autotrophic microphytobenthos (diatoms) in the food spectrum of these copepods. In contrast, the nematodes in hydrothermally active regions constantly showed depleted isotopic C and N signatures, which indicates their high dependence on chemosynthetic bacteria (see Sect. 1 in Chap. 5), a trophic specialization not found in nematodes from 'regular' deep-sea bottoms (Mordukhovich et al. 2018; Senokuchi et al. 2018).

Also in freshwater microcosm experiments, a close trophic interdependence of meiofauna and bacteria aggregated at the bottom surface was shown to exist (Liu et al. 2015). Not only were phototrophic bacterial biofilms the central food source for meiobenthos (see Sect. 1 in Chap. 2), their capacity to absorb nutrients from the environment was modulated by the bioturbative activity of meiofauna. Increased bioturbation caused a massive reduction in excess nitrogen and phosphate in the water. Conversely, nutrient addition to the water caused pulses of population growth in meiofauna (rotifers). These dynamics not only underline the interactive role of meiofauna and bacteria at the bottom interface, they also document the meiofaunal impact on water quality in rivers.

Also in marine sediments, the bioturbative activity of meiofauna has similarly positive effects (Bonaglia et al. 2014): bacteria-induced nitrogen fluxes (e.g. denitrification) and methane effluxes became massively stimulated by meiofauna activities. These results underline that sediment analyses for organic matter content alone do not suffice to describe the trophic status of a coastal area. It is only the integration of meiofauna into the compartments 'bacteria' and 'sediment' that results in robust conclusions on the environmental status (Bianchelli et al. 2016).

The closest interaction between meiobenthic organisms and bacteria has evolved where mutually positive effects favour symbiotic relationships. The general relevance of this evolutionary step culminates in the development of 'metaorganisms', the close combination of genetically different units (Bang et al. 2018). Multiple pathways towards symbioses are realized also among meiofauna. Here, they are not so rare and exclusive as originally thought. The repeated uptake of bacteria onto and into the bodies in some genera of interstitial marine nematodes and oligochaetes led to obligate symbioses. What is the driving factor in these evolutionary lines, how have these complex relationships arisen, how are they passed on? These studies not only offer chances for understanding novel evolutionary pathways (Sect. 3 in Chap. 6), they also lead to studies on life in extreme, nutrient-limited environments, on exotic metabolic pathways or environmental extremes—topics much noticed under scenarios of climate change.

Many stilbonematine nematodes living around the interface of oxic to sulphidic sediment layers are associated with several bacterial phylotypes of sulphur-oxidizing bacteria as ectosymbionts (Ott et al. 2004). The complex bacterial endosymbioses of two species-rich genera of tubificine oligochaetes are being analysed by the pioneering molecular work of Dubilier and her group. Their papers brought meiofauna to the front pages of high-profile journals (e.g. Dubilier et al. 2001, 2008; Zimmermann et al. 2016). Numerous publications reveal details of complex 'cyclotrophic' interactions between meiobenthic host, internal bacteria and the environment. They give insight into the functional and morphological substitution of usually obligatory organ systems (intestine, nephridia): the gutless worms completely depend on their incorporated bacteria.

The endobacterial consortium in these oligochaete species (some 20 spp. have been studied in genetical detail) consists of phylogenetically most diverse phylotypes: besides the more common lineages of gamma-, alpha- and delta-Proteobacteria, there is also one clade of Spirochaeta (Blazejak et al. 2005). Interestingly, the endosymbiotic gamma-Proteobacteria in the tubificine oligochaetes are closely related to those living as ectosymbionts in the stilbonematine nematodes mentioned before. This common occurrence at histologically different positions and in non-related animal phyla indicates multiple, evolutionary-independent steps, which apparently favour the hosts in their physiologically dynamic and demanding 'chemosynthetic' environments (Ruehland and Dubilier 2010; Zimmermann et al. 2016).

These ultrastructural, ecological and genetic analyses of symbiotic interactions between meiobenthos and bacteria, why are they examples of a most promising research pathway with a high future attention? It is the integration of general evolutionary aspects, revealed by diverse methods and under various ecological conditions, that leads to most progressive genetic studies and discovery of new metabolic pathways. All these fields represent approaches that, in combination with ecological data on distribution patterns and transmission pathways, can overcome difficulties inherent in work with tiny meiobenthic organisms. Putting meiofauna in a central hub position, these multifaceted studies at the (meio-)benthic level can help unravelling the little understood field of organismic interactions.

References

Avó AP, Daniell TJ, Neilson R et al (2017) DNA barcoding and morphological identification of benthic nematodes assemblages of estuarine intertidal sediments: advances in molecular tools for biodiversity assessment. Front Mar Sci 4:66. https://doi.org/10.3389/fmars.2017.00066

Baldrighi E, Manini E (2015) Deep-sea meiofauna and macrofauna diversity and functional diversity: are they related? Mar Biodiv 45(3):469–488. https://doi.org/10.1007/s12526-015-0333-9

Bang C, Dagan T, Deines P et al (2018) Metaorganisms in extreme environments: do microbes play a role in organismal adaptation? Zoology 127:1–19

Bellard C, Bertelsmeier C, Leadley P et al (2012) Impacts of climate change on the future of biodiversity. Ecol Lett 15:365–377

Bianchelli S, Pusceddu A, Buschi E, Danovaro R (2016) Trophic status and meiofauna biodiversity in the Northern Adriatic Sea: insights for the assessment of good environmental status. Mar Environ Res 113:18–30

Blazejak A, Erséus C, Amann R, Dubilier N (2005) Coexistence of bacterial sulfide-oxidizers, sulfate-reducers, and spirochetes in a gutless worm (Oligochaeta) from the Peru margin. Appl Environ Microbiol 71:1553–1561

Boeckner MJ, Sharma J, Proctor HC (2009) Revisiting the meiofauna paradox: dispersal and colonization of nematodes and other meiofaunal organisms in lowand high-energy environments. Hydrobiologia 624:91–106. https://doi.org/10.1007/s10750-008-9669-5

Boero F, Bonsdorff E (2007) A conceptual framework for marine biodiversity and ecosystem functioning. Mar Ecol 588:231–243

Bonaglia S, Nascimento FJA, Bartoli M et al (2014) Meiofauna increases bacterial denitrification in marine sediments. Nat Commun 5:5133

Bongers T (1990) The maturity index: an ecological measure of environmental disturbance based on nematode species composition. Oecologia 83:14–19

Brannock PM, Learman DR, Mahon AR et al (2018) Meiobenthic community composition and biodiversity along a 5500 km transect of Western Antarctica: a metabarcoding analysis. Mar Ecol Prog Ser 603:47-60. https://doi.org/10.3354/meps12717

Cerca J, Purschke G, Struck, TH (2018) Marine connectivity dynamics: clarifying cosmopolitan distributions of marine interstitial invertebrates and the meiofauna paradox. Mar Biol 165:123, pp 21

Costello MJ (2015) Biodiversity: the known, unknown, and rates of extinction. Curr Biol 25:R362–R383

Curini-Galletti M, Artois T, Delogu V et al (2012) Patterns of diversity in soft-bodied meiofauna: dispersal ability and body size matter. PLoS ONE 7(3):e33801. https://doi.org/10.1371/journal.pone.0033801

Danovaro R, Gambi C, Lampadariou N, Tselepides A (2008) Deep-sea nematode biodiversity in the mediterranean basin: testing for longitudinal, bathymetric and energetic gradients. Ecography 31:231–244. https://doi.org/10.1111/j.0906-7590.2008.5484.x

Danovaro R, Carugati L, Corinaldesi C et al (2013) Multiple spatial scale analyses provide new clues on patterns and drivers of deep-sea nematode diversity. Deep-Sea Res Pt. II-Top Stud Oceanogr 92:97–106. https://doi.org/10.1016/j.dsr2.2013.03.035

Derycke S, Meester ND, Rigaux A et al (2016) Coexisting cryptic species of the *Litoditis marina* complex (Nematoda) show differential resource use and have distinct microbiomes with high intraspecific variability. Mol Ecol 25:2093–2110

Dubilier N, Mülders C, Ferdelmann T et al (2001) Endosymbiotic sulphate-reducing and sulphide-oxidizing bacteria in an oligochaete worm. Nature 411:298–302

Dubilier N, Bergin C, Lott C (2008) Symbiotic diversity in marine animals: the art of harnessing chemosynthesis. Nat Rev Microbiol 6:725–740

Dümmer B, Ristau K, Traunspurger W (2016) Varying patterns on varying scales: a metacommunity analysis of nematodes in European lakes. PLoS ONE 11(3):e015 1866. https://doi.org/10.1371/journal.pone.0151866

Fonseca VG, Carvalho GR Sung W et al (2010) Second-generation environmental sequencing unmasks marine metazoan biodiversity. Nat Commun 1:98. https://doi.org/10.1038/ncomms1095

Fonseca VG, Carvalho GR, Nichols B et al (2014) Metagenetic analysis of patterns of distribution and diversity of marine meiobenthic eukaryotes. Global Ecol Biogeogr 23:1293–1302

Fonseca G, Fontaneto D, Di Domenico M (2018) Addressing biodiversity shortfalls in meiofauna. J Exp Mar Biol Ecol 502:26–38 knowlhttp://dx.doi.org/10.1016/j.jembe.2017.05.007

Gansfort B, Uthoff J, Traunspurger W (2018) Interactions among competing nematode species affect population growth rates. Oecologia 187:75–84

Giere O (2009) Meiobenthology. The microscopic motile fauna of aquatic sediments, 2nd edn. Springer, Berlin Heidelberg, pp 527

Hohberg K, Traunspurger W (2009) Foraging theory and partial consumption in a tardigrade-nematode system. Behav Ecol 20:884–890

Leasi F, Da Silva SC, Norenburg J (2016) At least some meiofaunal species are not everywhere. Indication of geographic, ecological and geological barriers affecting the dispersion of species of *Ototyphlonemertes* (Nemertea, Hoplonemertea). Mol Ecol 25:1381–1397

Leasi F, Sevigny JL, Laflamme EM et al (2018) Biodiversity estimates and ecological interpretations of meiofaunal communities are biased by the taxonomic approach. Nat Commun Biol 1:112. https://doi.org/10.1038/s42003-018-0119-2

Leduc D, Probert PK, Duncan A (2009) A multi-method approach for identifying meiofaunal trophic connections. Mar Ecol Prog Ser 383:95–111

Leduc D, Rowden AA, Pilditch CA et al (2013) Is there a link between deep-sea biodiversity and ecosystem function? Mar Ecol 34:334–344. https://doi.org/10.1111/maec.12019

Lee MR, Riveros M (2012) Latitudinal trends in the species richness of free-living marine nematode assemblages from exposed sandy beaches along the coast of Chile (1842TS). Mar Ecol 33:317–325

Lins L, Leliaert F, Riehl T et al (2017) Evaluating environmental drivers of spatial variability in free-living nematode assemblages along the Portuguese margin. Biogeosciences 14:651–669. https://doi.org/10.5194/bg-14-651-2017

Liu Y, Majdi N, Tackx M et al (2015) Short-term effects of nutrient enrichment on river biofilm: $N-NO_3$- uptake rate and response of meiofauna. Hydrobiologia 744:165–175

Maghsoud H, Weiss A, Smith J III et al (2014) Diagnostic PCR can be used to illuminate meiofauna diets and trophic relationships. Invert Biol 133:121–127

Maidi N, Traunspurger W (2017) Leaf fall affects the isotopic niches of meiofauna and macrofauna in a stream food web. Food Webs 10:5–14

Mendes CB, Norenburg JL, Solferini VN, Andrade SC (2018) Hidden diversity: phylogeography of genus *Ototyphlonemertes* Diesing, 1863 (Ototyphlonemertidae: Hoplonemertea) reveals cryptic species and high diversity in Chilean populations. PLoS ONE 13(4):e0195833

Mialet B, Majdi N, Tackx M et al (2013) Selective feeding of bdelloid rotifers in river biofilms. PLoS ONE 8(9):e75352. https://doi.org/10.1371/journal.pone.0075352

Moens T, De Meester N, Guilini K et al (2016) Experimental elucidation of meiofauna trophic interactions: from radioactive tracer techniques to next generation sequencing. ISIMCO 16th Intern Meiofauna Confer, Book of Abstracts: 6–7 ISBN 978-980-9798-29-7

Mordukhovich VV, Kiyashko SI, Kharlamenko VI, Fadeeva NP (2018) Determination of food sources for nematodes in the Kuril Basin and eastern slope of the Kuril Islands by stable isotopes and fatty acid analyses. Deep Sea Res Pt II. https://doi.org/10.1016/j.dsr2.2018.01.003

Muschiol D, Giere O, Traunspurger W (2015) Population dynamics of a cavernicolous nematode community in a chemoautotrophic groundwater system. Limnol Oceanogr 60:127–135

Ott J, Bright M, Bulgheresi S (2004) Marine microbial thiotrophic ectosymbioses. Oceanogr Mar Biol Ann Rev 42:95–118

Packmor J, Riedl T (2016) Records of Normanellidae Lang, 1944 (Copepoda, Harpacticoida) from Madeira Island support the hypothetical role of seamounts and oceanic islands as "stepping stones" in the dispersal of marine meiofauna. Mar Biodivers 46:861–877. https://doi.org/10.1007/s12526-016-0448-7

Patrício J, Adão H, Neto JM et al (2012) Do nematode and macrofauna assemblages provide similar ecological assessment information? Ecol Indicat 14:124–137

Pusceddu A, Carugati L, Gambi C et al (2016) Organic matter pools, C turnover and meiofauna biodiversity in the sediments of the western Spitsbergen deep continental margin, Svalbard Archipelago. Deep-Sea Res Pt I Oceanogr Res Pap 107:48–58. https://doi.org/10.1016/j.dsr.2015.11.004

Rosli N, Leduc D, Rowden AA, Probert PK (2017) Review of recent trends in ecological studies of deep-sea meiofauna, with focus on patterns and processes at small to regional spatial scale. Mar Biodiv 48:13–34. https://doi.org/10.1007/s12526-017-0801-5

Ruehland C, Dubilier N (2010) Gamma- and epsilon proteobacterial ectosymbionts of a shallow-water marine worm are related to deep-sea hydrothermal vent ectosymbionts. Environ Microbiol 12:2312–2326. https://doi.org/10.1111/j.1462-2920.2010.02256.x

Schmid-Araya JM, Schmid PE, Tod SP, Esteban GF (2016) Trophic positioning of meiofauna revealed by stable isotopes and food web analyses. Ecology 97:3099–3109

Semprucci F, Colantoni P, Balsamo M (2016) Is maturity index an efficient tool to assess the effects of the physical disturbance on the marine nematode assemblages? A critical interpretation of disturbance-induced maturity successions in some study cases in Maldives. Acta Oceanol Sin 35:89–98

Senokuchi R, Nomaki H, Watanabe HK (2018) Chemoautotrophic food availability influences copepod assemblage composition at deep hydrothermal vent sites within sea knoll calderas in the northwestern Pacific. Mar Ecol Prog Ser 607:37–51

Stoch F, Artheau M, Brancelj A (2009) Biodiversity indicators in European ground waters: towards a predictive model of stygobiotic species richness. Freshw Biol 54:745–755

Turner EL, Montagna PA (2016) The max bin regression method to identify maximum bioindicator responses to ecological drivers. Ecol Informat 36:118–125

Vanaverbeke J, Merckx B, Degraer S, Vincx M (2011) Sediment-related distribution patterns of nematodes and macrofauna: two sides of the benthic coin? Mar Environ Res 71:31–40. https://doi.org/10.1016/j.marenvres.2010.09.006

Wieser W (1953) Die Beziehung zwischen Mundhöhlengestalt, Ernährungsweise und Vorkommen bei freilebenden marinen Nematoden. Ark Zool 2:439–484

Zimmermann J, Wentrup C, Sadowski M et al (2016) Closely coupled evolutionary history of ecto- and endosymbionts from two distantly related animal phyla. Mol Ecol 25:3203–3223

Chapter 5
Physiology, Biochemistry and Meiofauna—A Rarely Touched Terrain

Interpreting the 'cause and effect' sequence physico-chemically: Great potential but rarely applied in meiobenthology

Ecophysiological reactions of organisms and the metabolic pathways involved are and will remain a cornerstone in general biological research. In the realm of meiobenthos, however, physiological studies are still scarce, often even absent. At the last International Meiofauna Conference (Crete 2016), just three out of some 160 contributions focussed on physiological and ecophysiological topics!

In the old days, this scarcity of publications was understandable: lack of sufficient sample material and related technical problems. But today's modern methods, such as immunofluorescence microscopy visualizing immunoreactive reactions, radioactive marking (see Sect. 1 in Chap. 6) plus molecular genetics, allow biochemical and physiological insights on a cellular or even molecular level. Nano-scale elemental and isotopic analyses (NanoSIMS), when connected with electron microscopy or atomic force microscopy, allow for tracking of biochemically relevant molecules such as peptides or lipids (Jiang et al. 2016). The detailed molecular pathways involved can be visualized by application of specific, new probes. But, as shown above, in meiobenthology, a strong bias against physiological studies seems to remain even today.

Discussing physiological aspects of ocean acidification, Pörtner (2008) gave valuable directives applicable also for future meiobenthos research: a more accurate assessment of quantitative scenarios in ecosystem research requires a 'cause and effect understanding' beyond empirical observation. Physico-chemical factors influence animals indirectly via physiological reactions such as ion transport systems. Therefore, environmental recordings 'are useful proxies but usually not direct drivers'. This conclusion not only underlines the need of physiological investigations also in meiobenthos, it also stands for the promising potential of these studies. When analysing the impacts of environmental factors, it is physiology that can disentangle the underlying factorial complexes and helps to understand the frequently synergistic actions.

O. Giere, *Perspectives in Meiobenthology*, SpringerBriefs in Biology, https://doi.org/10.1007/978-3-030-13966-7_5

The strength of accurate physiological recordings versus ecological observations/interpretations is well demonstrated in the oil pollution study by Shchapova et al. (2018). Although not dealing with meiofauna, this paper is remarkable: rather than ascertaining the usual survival/mortality relations after exposure to crude oil, the authors evaluate the physiological reaction of sensitive species (amphipods) to toxic stress by measuring the concentrations of haemolymphic stress-response markers. Even at low oil concentrations, so far considered ecologically safe, markers such as heat-shock proteins (HSP 70), lipid peroxidation rates and specific enzymes were clearly impacted by oil. These quantitative results have the potential of questioning established standards, and they can help to set new threshold limits. Similar studies urgently need repetition, this time with meiofauna and in other oil-contaminated regions!

The tiny amounts of material required today for physiological studies should set aside most technical limitations. Hence, I contend, the reasons for the scarcity of physiological research in meiobenthology are more historical than technical ones. Historically, the young discipline 'meiobenthology' was based on morphology, taxonomy and aspects of distribution (compare Giere 2009: Chap. 1). Excursions into field ecology followed, while questions for the physiological background of processes and functions in meiofaunal ecosystems stood long behind and were often prompted by ecological extremes.

Today, it seems: we are often describing effects of substances and reactions on meiofauna, but what is the reason for these effects and which pathways are involved? These are not trivial questions, and their answers would give access to a new understanding of environmental problems. Here, I present four fields, which should entail promising approaches for future physiological work, again underlining the potential of working with meiobenthos.

1 Hypoxia, Anoxia and Hydrogen Sulphide—Fields of Challenge

Bulk respiration rates misleading – Oxygen not indispensible for meiofaunal life – Role of bacteria as partners

Reduced oxygen supply, often combined with toxic hydrogen sulphide, represents one of those extreme situations that increasingly challenge benthic life in our oxic world. Today, with globally rising temperatures, hypoxic areas are spreading. Although small invertebrates with their low compensation potential for extracellular disturbances are usually seen as disadvantaged compared to macrofauna, the capacity of many meiobenthic forms to tolerate suboxic or even temporarily anoxic conditions is amazing. Hence, clarifying the metabolic pathways involved is one of the chances to discover new physiological strategies. They could receive considerable scientific influence under the predicted climate scenarios.

With oxygen as one of the potent structuring biofactors in metazoan communities, oxygen consumption rates (respiration) are often considered a convenient basis for calculating biomass. However, as shown by Braeckman et al. (2013), the metabolic reactions to oxygen are highly variable in meiofauna and respiration rates often depend on environmental characteristics. Therefore, measuring gross respiration rates of samples as rather invariable, non-interacting parameters tends to misinterpret the meiofaunal role. Only when those physiological processes are considered that determine variability of respiration, can we correctly derive biomass from respiration rates and then calculate the meiofaunal contribution to an overall benthic carbon cycling.

In many habitats, low oxygen concentrations do not allow for a normoxic metabolism of meiofauna. Active migration to micro-oxic niches is one reaction, and another is a flexible metabolism temporarily withstanding hypoxic or even anoxic conditions. Many meiobenthic animals (especially among ciliates, gastrotrichs, kinorhynchs, gastrotrichs and nematodes) are good candidates for this metabolic versatility (for details, see Giere 2009). In the most species-rich taxon, the nematodes, many representatives can even survive longer-lasting anoxia (Steyaert et al. 2007; Braeckmann et al. 2013; Muschiol et al. 2015; Van Campenhout et al. 2016). Hence, analyses of the biochemical pathways underlying conditions of oxygen limitation, as well as the molecular and cellular mechanisms involved, should become a prime topic when studying physiological reactions of meiobenthos. At least, they should refine assessment of bulk parameters such as oxygen consumption, n50-survival rates or avoidance reactions.

Recent sampling at the Black Sea bottom gave new evidence that under hypoxia, specialized meiofauna can constantly live and reproduce (Sergeeva and Zaika 2013). Specimens from several meiobenthic taxa have even been found to live in complete anoxia (Giere 2009, Chap. 8.4). How can this meiofauna overcome the physiological constraints of permanent anoxia?—a much discussed problem with relevance for general metazoan life and its origin. Are there Eumetazoa that not only survive anoxic periods but thrive in constant and full anoxia?

As mentioned in the Introduction, some loriciferans in various reproductive stages were found in permanently anoxic basins of the Mediterranean Sea (Danovaro et al. 2010). Why is this outstanding study mentioned here again? It meticulously combines results from biochemical analyses, radioactive tracing, sophisticated optical analyses and phylogenetic considerations. These together give insights into aberrant biochemical pathways and document 'compelling evidence' that 'these metazoans live under anoxic conditions through an obligate anaerobic metabolism that is similar to that demonstrated so far only for unicellular eukaryotes'. The acristate mitochondria of these loriciferans are hydrogenosomes, anaerobic derivates of mitochondria, which generate hydrogen as a by-product of ATP production by pyruvate oxidation (Fig. 1). This pathway, so far novel to free-living animals, is known from some flagellates, ciliates and fungi living under anaerobic conditions (see Mentel and Martin 2010; Müller et al. 2012). Not surprisingly, the report on this meiofauna with unique metabolic routes reached the front pages of top journals and even popular newspaper articles. We definitely need further organismic and molecular genetic studies on

Fig. 1 Electron micrograph of hydrogenosome with indicative 'marginal plate' (arrow) in tissue of *Rugiloricus* (Loricifera) from an anoxic hypersaline basin of the Mediterranean; scale bar 0.2 micrometers (from: Danovaro et al. 2010, modified; background in left upper corner retouched)

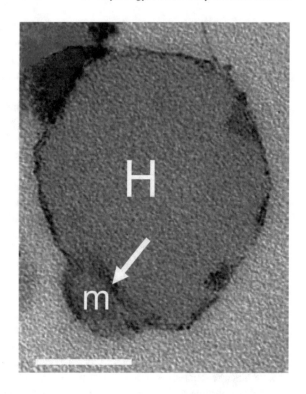

those metazoans occurring under oxygen-limited conditions and their physiologically enigmatic pathways. The general relevance of this research for considerations on the rise of metazoan life will bring meiofauna in the tracks towards general biology research.

In many sediments, hypoxia/anoxia is linked to toxic hydrogen sulphide. Sulphide stress enhancing oxidative stress renders the existence of animals with 'normal' oxidative metabolism impossible. But for a specifically adapted meiofauna, sulphide does not mean exclusion from habitable conditions. Sediment samples from oxygen-minimum zones, mud volcanoes, pock marks and seeps, both hot and cold, document the existence of a rich and diverse meiofauna. Many of these areas are exposed to hydrogen sulphide in concentrations prohibiting 'normal' animal life. And yet, cold seeps have been contended as 'islands of primary production for … meiofaunal communities' Van Gaever et al. (2009). Also in the Mediterranean Sea, mud volcanoes were ranked as 'hot spots of exclusive meiobenthic species' (Zeppilli et al. 2011). Their biocoenoses were characterized by a restricted occurrence of rare species, low connectivity and high rates of endemism. Sometimes oxygen-depleted sites even harboured more abundant meiofauna than fully oxic sediments (Veit-Köhler et al. 2009).

Nematodes are the prevailing taxon in chemosynthetic habitats and oxygen-reduced waters (Van Gaever et al. 2009). However, the physiological (and distributional) adaptations enabling this phylum to colonize these hostile and isolated deep-sea environments are largely unknown. Are they specifically adapted or just living at their extreme range of tolerance taking advantage of reduced competition? Around hot deep-sea vents, also harpacticoids of the family Dirivultidae seem to thrive in the critical conditions of venting fluids (see Plum et al. 2017).

Also in more shallow sulphidic sites, nematodes are regularly found to populate the sediments. Dahms et al. (2018) pointed to the role of shallow-water thermal vents (near Taiwan) as easily accessible study areas inviting for in situ studies and long-term experiments. The authors even compared these sites as templates for global change scenarios (high temperatures, increased concentrations of CO_2 and toxicants).

In shallow sites, some of the 'thiobiotic' specialists harbour a complex pattern of symbioses with one or several groups of bacteria (for details see Sect. 3 in Chap. 4). Discovered in some meiobenthic nematodes and oligochaetes, molecular genetic analyses of the hosts and their bacterial partners not only led to an understanding of complex and novel metabolic interactions, they also allowed first conclusions on the evolutionary pathways involved. The close relationship between the ectosymbiotic bacteria living on the cuticle of nematodes and the endosymbionts under the cuticle of oligochaetes from divergent regions of the sea remains surprising. It indicates a sustained process in these specialized meiobenthic symbioses of incorporating apparently widely distributed bacteria (Zimmermann et al. 2016; see Fig. 2). Generalizing this aspect, extreme environments seem to favour complex synergistic, multi-organismic functional units, in this case eukaryotic hosts and microbial communities, to form meta-organisms (Bang et al. 2018).

2 Temperature—A Principal Physiological Driver

Cryobiosis: life below freezing point – Role of bioprotectants and anti-freeze proteins

With the growing focus on polar regions, the survival mechanisms in animals, also meiobenthos, under extremely cold conditions will gain new relevance. Rather early have the Tardigrada taken their reputation as the champions of the cold. Only recently, their potential for cryobiosis was widely reported in the media when their resting stages became exposed to outer space conditions where they survived in 'a reversible, ametabolic state'. The subsequent metabolic analyses, compiled by Møbjerg et al. (2011), point to the intensive synthesis of 'bioprotectants' after desiccation. These consist of special carbohydrates (e.g. trehalose) and heat-shock proteins as well as antioxidative enzymes plus other compounds scavenging free radicals. However, this would not explain the potential of tardigrades in their active stage to even survive temperatures of −20 °C and additional environmental extremes. Here, probably a special hyperosmotic potential combined with highly efficient DNA repair processes is involved.

Some (calanoid) copepods thrive in the channel system of sea ice at extreme salinities and variable temperatures. The study by Kiko (2010) emphasizes the phys-

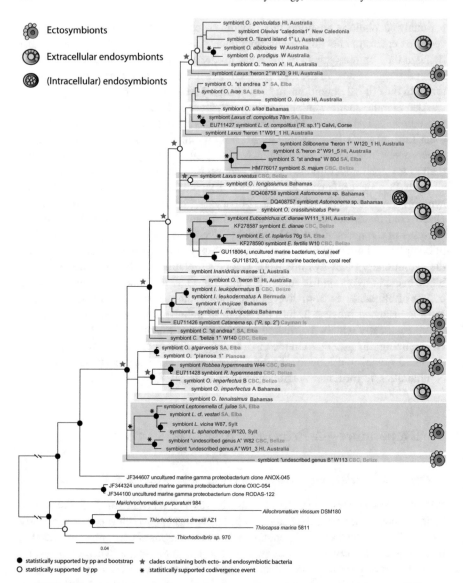

Ectosymbionts

Extracellular endosymbionts

(Intracellular) endosymbionts

● statistically supported by pp and bootstrap   ★ clades containing both ecto- and endosymbiotic bacteria
○ statistically supported by pp   * statistically supported codivergence event

Fig. 2 Relationship of bacteria occurring as symbionts in marine oligochaetes (internally) and on nematodes (externally). This speaks for an underlining independent and parallel evolution of symbioses in these groups (from: Zimmermann et al. 2016, modified)

iological impact of a surrounding with strongly fluctuating minus temperatures. The copepods lower the freezing point of their cell fluids through thermal hysteresis of special proteins in their haemolymph. These antifreeze proteins bind to the surface of ice crystals, preventing crystallization of body fluids and stabilizing the cell membranes. The specific molecular structure of the antifreeze proteins in these tiny copepods contrasts with that of better-studied polar fish and seems unique in metazoans. On the other hand, the author found a strong homology to various polar diatoms and bacteria. This parallel invites for expansion of these physiological studies also to sympagic meiofauna in polar areas and in sea ice where foraminiferans, nematodes, harpacticoids and turbellarians frequently occur (see Sect. 3 in Chap. 2).

Beyond survival under extreme conditions, temperature is probably the most powerful driver of general physiological processes such as growth and fecundity. Is the classical temperature-size rule (TSR: individuals maintained at low temperatures grow more slowly, but attain a larger size upon maturity) valid in meiobenthic species that are found both in temperate and polar regions?—an open field for experimental studies with meiobenthos that would not only allow for conclusions on the evolution of adaptive physiological pathways under zoogeographical aspects, but also gain its relevance under the auspices of climate change.

3 Fatty Acids and Stable Isotopes as Biomarkers

Fatty acids reveal food sources and trophic steps — Their variability as a valuable indicator of trophic changes – Stable isotopes monitor carbon flow

Only few publications on meiofauna are found to directly measure the biochemical profiles of essential metabolic substances, e.g. of biomarkers involved in vital functions of organisms. Among these biomarkers, unsaturated fatty acids (UFAs) attain a predominant role in conveying energy from primary producers to higher trophic levels. In all animals, they are essential for any cell formation, for lipid metabolism and health (Filimonova et al. 2016, and for meiofauna: Werbrouck et al. 2017). Hence, their composition and content can be used as sensitive qualitative and often quantitative markers of trophic pathways in meiobenthic organisms, especially primary consumers. As temperature seems to affect uptake, metabolism and composition of polyunsaturated fatty acids (PUFAs), the predicted rise in ocean temperatures is depleting fatty acid content in phytoplankton. But not only at the basis of the food web seem rising temperatures to alter the fatty acid content and metabolism, an aspect rarely considered in general discussions (Gourtay et al. 2018). Consequently, analyses of fatty acid profiles, often in connection with stable-isotope relations, can unravel essential food webs giving valuable insights into pathways in biochemical ecology. But they also allow for conclusions on environmental changes in trophic chains.

Those PUFAs that are restricted to either bacteria, or suspended particles, or higher plants attain a particularly indicative relevance: they allow for conclusions on their role in meiofauna nutrition and indicate metabolic pathways (Cnudde et al. 2015). Detailed analyses of fatty acids combined with an assessment of carbon signatures allowed Van Gaever et al. (2009) to differentiate the trophic sources of dominant meiofauna species around a cold seep mud volcano: dominance of characteristic bacterial biomarkers and exceptionally light carbon signatures characterized the prevailing harpacticoids near the volcano's centre as utilizer of methanotrophic bacteria. In contrast, nematodes, common in the neighbouring microbial mats, showed marked differences: their signatures indicated a food spectrum based on sulphide-oxidizing bacteria. A corresponding trophic difference in specific bacterial fatty acids was found in annelids around the oxic/sulphidic boundary layer. It depended on the worms' preferred occurrence in anoxic or oxic sediments (Giere et al. 1991). Also, contamination and subsequent bioaccumulation of pesticides and heavy metals has been revealed by analysing profiles and concentrations of essential fatty acids (Filimonova et al. 2016).

Among meiofauna, many harpacticoid copepods attain an intermediate position between autotrophic and heterotrophic pathways. Grazing on diatoms, they import plant-derived dietary lipids into the food chain. Hence, they are in the focus of studies on fatty acid composition and change during food chain transport. Experimenting with the intertidal harpacticoid *Platychelipus littoralis*, Cnudde et al. (2015) demonstrated the relation and variability of PUFA profiles under variable temperatures: with rising temperatures, important storage fatty acids became depleted, indicating signs of metabolic stress. Longer or more extreme heat stress lowered the concentration of the omega-3-fatty acid DHA implying negative nutritional effects, a gloomy perspective for global warming (Werbrouck et al. 2016). Moreover, under insufficient or low-quality food, the synthesis of essential fatty acids is changed. This underlines the potential of these physiological experiments as indicators of environmental stress (Werbrouck et al. 2017).

Only recently, a most promising new approach to elucidate the flow of carbon from bacteria up to meiobenthic consumers (experiments with *Escherichia coli* and *Caenorhabditis elegans*) has been developed: combining stable-isotope signals and Raman spectroscopy 'offers new insights … and provides a novel research tool in the investigation of biochemical pathways' (Li et al. 2013; Wang et al. 2016). A label-free and non-destructive method, this approach could even reveal the function of single microorganisms in their natural environment.

4 Physiological Pathways Revealed by the Genetic Background

The molecular-genetic basis rather than biochemical analyses – Physiological conclusions from 'metabolomics'

Most physiological/biochemical investigations using meiobenthos as study object are complicated and highly theory-based. This may be a reason why analyses of the underlying genetic background might be preferable as indirect substitutes of the resulting biochemical steps. Today, the increasingly comprehensive knowledge of the molecular genetic basis allows for reconstruction of the metabolic pathways involved. State-of-the-art analysis (e.g. transcriptomes, protein-encoding genes) can often identify most complex coding pathways and explain extraordinary physiological capabilities and ecological adaptations. Thus, the traditional difficulties in revealing details of biochemical steps and physiological reactions are circumvented by analysing the coding genetic background.

A good example from meiobenthos is the transcriptome analysis of two cryptic nematode species, *Halomonhystera* GD 1, living in intertidal sediments, and *Halomonhystera hermesi*, living in mud volcanoes in the deep sea (Van Campenhout et al. 2016): in the deep-sea species, expression of those genes coding for oxidative phosphorylation and fatty acid metabolism is stronger than in the morphologically identical intertidal species. Also in other genes differences were found: expression only in the deep-sea species and inactivity in the intertidal one. In contrast, in the intertidal *H. hermesi*, those gene activities were enhanced that coded for rapidly signalling and adaptive transcriptional programming. The authors interpret these results as an adaptation to the respective biotopic conditions (deep sea: static, slow change, hypoxic vs. intertidal: highly dynamic, oxic). In this group of cryptic species, the high plasticity of their genetic background explains why extreme habitats, both in the deep sea and in shallow zones, can be colonized. However, the underlying physiological steps remain, as yet, unresolved.

Another example where the analysis of detailed physiological pathways and mutual interactions is primarily based on comprehensive molecular genetics: the study of gutless meiobenthic annelids characterized by complex and obligate symbioses with various bacteria. In the interstitial marine oligochaete *Olavius algarvensis,* metaproteomic and metabolomic analyses could show that novel symbiotic bacterial consortia enabled physiological pathways previously unknown from most eukaryotes (see Blazejak et al. 2005; Dubilier et al. 2008; Zimmermann et al. 2016; Bergin et al. 2018). Just one of them is the oxidation of carbon monoxide using hydrogen as an energy source (Kleiner et al. 2015), a metabolic pathway previously unknown from free-living animals.

References

Bang C, Dagan T, Deines P et al (2018) Metaorganisms in extreme environments: do microbes play a role in organismal adaptation? Zoology 127:1–19

Bergin C, Wentrup C, Brewig N et al (2018) Acquisition of a novel sulfur-oxidizing symbiont in the gutless marine worm *Inanidrilus exumae*. Appl Environ Microbiol 84:e0226717. https://doi.org/10.1128/AEM.02267-17

Blazejak A, Erséus C, Amann R, Dubilier N (2005) Coexistence of bacterial sulfide-oxidizers, sulfate-reducers, and spirochetes in a gutless worm (Oligochaeta) from the Peru margin. Appl Environ Microbiol 71:1553–1561

Braeckman U, Vanaverbeke J, Vincx M et al (2013) Meiofauna metabolism in suboxic sediments: currently overestimated. PLoS ONE 8:e59289

Cnudde C, Moens T, Werbrouck E et al (2015) Trophodynamics of estuarine intertidal harpacticoid copepods based on stable isotope composition and fatty acid profiles. Mar Ecol Prog Ser 524:225–239

Dahms H-U, Schizas NV, James RA et al (2018) Marine hydrothermal vents as templates for global change scenarios. Hydrobiologia 818:1–10. https://doi.org/10.1007/s10750-018-3598-8

Danovaro R, Dell'Anno A, Pusceddu A et al (2010) The first metazoa living in permanently anoxic conditions. BMC Biol 8:30–40

Dubilier N, Bergin C, Lott C (2008) Symbiotic diversity in marine animals: the art of harnessing chemosynthesis. Nature Rev Microbiol 6:725–740

Filimonova V, Gonçalves F, Marques JC et al (2016) Fatty acid profiling as bioindicator of chemical stress in marine organisms: a review. Ecol Indicat 67:657–672

Giere O (2009) Meiobenthology. The microscopic motile fauna of aquatic sediments, 2nd edn. Springer, Berlin, Heidelberg, 527 pp

Giere O, Conway NM, Gastrock G, Schmidt C (1991) 'Regulation' of gutless annelid ecology by endosymbiotic bacteria. Mar Ecol Prog Ser 68:287–299

Gourtay C, Chabot D, Audet C et al (2018) Will global warming affect the functional need of essential fatty acids in juvenile sea bass (*Dicentrarchus labrax*)? A first overview of the consequences of lower availability of nutritional fatty acids on growth performance. Mar Biol 165:143. https://doi.org/10.1007/s00227-018-3402-3

Jiang H, Kilburn MR, Decelle J, Musat N (2016) NanoSIMS chemical imaging combined with correlative microscopy for biological sample analysis. Curr Opin Biotechnol 41:130–135

Kiko R (2010) Acquisition of freeze protection in a sea-ice crustacean through horizontal gene transfer? Polar Biol 33:543–556

Kleiner M, Wentrup C, Holler C et al (2015) Use of carbon monoxide and hydrogen by a bacteria-animal symbiosis from seagrass sediments. Environ Microbiol 17:5023–5035

Li M, Huang WE, Gibson CM et al (2013) Stable isotope probing and Raman spectroscopy for monitoring carbon flow in a food chain and revealing metabolic pathway. Anal Chem 85:1642–1649

Mentel M, Martin W (2010) Anaerobic animals from an ancient, anoxic ecological niche. BMC Biol 8:32. https://doi.org/10.1186/1741-7007-8-32

Møbjerg N, Halberg KA, Jørgensen A et al (2011) Survival in extreme environments—on the current knowledge of adaptations in tardigrades. Acta Physiol 202:404–420

Müller M, Mentel M, Van Hellemond JJ et al (2012) Biochemistry and evolution of anaerobic energy metabolism in eukaryotes. Microbiol Mol Biol Rev 76:444–495

Muschiol D, Giere O, Traunspurger W (2015) Population dynamics of a cavernicolous nematode community in a chemoautotrophic groundwater system. Limnol Oceanogr 60:127–135

Plum C, Pradillon F, Fujiwara Y et al (2017) Copepod colonization of organic and inorganic substrata at a deep-sea hydrothermal vent site on the Mid-Atlantic ridge. Deep-Sea Res Part II Topical Stud 137:335–348. https://doi.org/10.1016/j.dsr2.2016.06.008

Pörtner H-O (2008) Ecosystem effects of ocean acidification in times of ocean warming: a physiologist's view. Mar Ecol Prog Ser 373:203–217

Sergeeva N, Zaika V (2013) The Black Sea meiobenthos in permanently hypoxic habitat. Acta Zool Bulg 65:139–150

Shchapova EP, Axenov-Gribanov DV, Lubyaga YA et al (2018) Crude oil at concentrations considered safe promotes rapid stress-response in Lake Baikal endemic amphipods. Hydrobiologia 805:189–201. https://doi.org/10.1007/s10750-017-3303-3

Steyaert M, Moodley L, Nadong T et al (2007) Responses of intertidal nematodes to short-term anoxic events. J Exp Mar Biol Ecol 345:175–184

Van Campenhout J, Vanreusel A, Van Belleghem S, Derycke S (2016) Transcription, signaling receptor activity, oxidative phosphorylation, and fatty acid metabolism mediate the presence of closely related species in distinct intertidal and cold-seep habitats. Genom Biol Evol 8:51–69

Van Gaever S, Moodley L, Pasotti F et al (2009) Trophic specialisation of metazoan meiofauna at the Håkon Mosby mud volcano: fatty acid biomarker isotope evidence. Mar Biol 156:1289–1296

Veit-Köhler G, Gerdes D, Quiroga E et al (2009) Metazoan meiofauna within the oxygen-minimum zone off Chile: Results of the 2001-PUCK expedition. Deep Sea Res Part II: Topical Stud 56:1105–1111. https://doi.org/10.1016/j.dsr2.2008.09.013

Wang Y, Huang WE, Cui L, Wagner M (2016) Single cell stable isotope probing in microbiology using Raman microspectroscopy. Curr Opin Biotechnol 41:34–42

Werbrouck E, Van Gansbeke D, Vanreusel A et al (2016) Temperature-induced changes in fatty acid dynamics of the intertidal grazer *Platychelipus littoralis* (Crustacea, Copepoda, Harpacticoida): insights from a short-term feeding experiment. J Therm Biol 57:44–53

Werbrouck E, Bodé S, Van Gansbeke D et al (2017) Fatty acid recovery after starvation: insights into the fatty acid conversion capabilities of a benthic copepod (Copepoda, Harpacticoida) Mar Biol 164: 151 https://doi.org/10.1007/s00227-017-3181-2

Zeppilli D, Mea M, Corinaldesi C, Danovaro R (2011) Mud volcanoes in the mediterranean Sea are hot spots of exclusive meiobenthic species. Prog Oceanogr 91:260–272

Zimmermann J, Wentrup C, Sadowski M et al (2016) Closely coupled evolutionary history of ecto- and endosymbionts from two distantly related animal phyla. Mol Ecol 25:3203–3223. https://doi.org/10.1111/mec.13554

Chapter 6
Towards an Integrated Triad: Taxonomy, Morphology and Phylogeny

Traditional fields updated and strengthened by novel technical solutions – Distribution patterns under scrutiny – Novel structural analyses reveal unknown details – Often far-reaching results for phylogeny – Comparison to world-wide archives needed

In the early days of meiobenthology (this term did not even exist), researchers were confronted with a plethora of new taxa and novel structures, from species to phylum. Thus, they focussed on taxonomic descriptions and morphological studies. But today, can these traditional domains still offer something 'new under the sun'? Looking at the amazingly fruitful recent literature, this question is easily answered. Computer analyses help to construct molecular genetic patterns, not only of some specimens, oftentimes of whole populations or even the entire bulk of meiofauna contained in sample sets. This allows for tackling taxonomic problems in a previously unknown dimension: the genetic background of morphologically indiscernible intraspecific variation can become revealed (Pereira et al. 2010). This, in turn, allows for detailed interpretations of complex distribution patterns.

Many meiofauna taxa are known for their cosmopolitan distribution. Is this unusual pattern real or rather based on traditional morphology not considering genetic variation below the morphological level of expression? In many cases, a differentiated analysis of genes would lead to a corrected distribution picture (see Sect. 2 in Chap. 4). Today, the degree of cryptic diversity in flocks of homomorphous species can be proven and the extent of sympatry realistically assessed (Derycke et al. 2010; Karanovic and Kim 2014; Grosemans et al. 2016). On the other hand, overestimated species numbers and biodiversity estimates can be corrected by elimination of synonyms (Costello 2015).

New preparative procedures, staining methods and novel microscopic developments (e.g. immuno-histochemistry, fluorescence) integrated with digital evaluation and computer-based analysis reveal previously indiscernible anatomical structures, often with surprisingly new taxonomic and phylogenetic consequences. Finally, comparison of singular results with the stocks of corresponding data stored in huge electronic databases allows for scrutiny even at remote places. This, in turn, enables evaluation, both taxonomical and ecological, to a formerly unknown extent. Com-

© The Author(s), under exclusive license to Springer Nature Switzerland AG 2019
O. Giere, *Perspectives in Meiobenthology*, SpringerBriefs in Biology,
https://doi.org/10.1007/978-3-030-13966-7_6

pared to macrobenthos, meiobenthos with its high population numbers per sample can often provide fairly reliable data sets. Hence, this 'future' meiobenthology, interactively linking digital and traditional progress, will be able not only to substantially contribute to a deepened knowledge. It will gain relevance to overall benthology and even to general biology.

1   Advances at the Methodological Basis

Pitfalls and caveats of gene taxonomy – Broad genetic basis (-omics) will reduce errors – A promising mass-spectrometric method based on proteome fingerprints – Museum collections and morphospecies still required – Meiofauna in large data banks

Even basic methodological standards of meiofauna work, both in the field and in the laboratory, such as sample coring, fauna extracting or data evaluating, have been improved and are subject to progress unimagined in earlier years. Conveniently foldable field corers, specialized deep-sea grabs or even ROV/AUV-mounted remotely controlled corers for deep-water sediments are available now (Ocean Scientific International Ltd., OSIL, UK). These days, effective, even autonomous sampling, application of high-throughput analyses under computer control and semi-automated identification (basing on principles of modern clinical cytometers and often used in plankton research) enable evaluation of large sample series as required for forthcoming impact studies (compare Xu et al. 2011; Kitahashi et al. 2018). In this context, it is worthwhile noting that the classical Ludox centrifugation as a convenient separation method will retain its relevance even in the future since it does not interfere with subsequent molecular tools such as DNA extraction and amplification (Fonseca et al. 2011).

Confocal microscopy can depict fluorescence-labelled meiofauna in undisturbed cores and show their distribution/migration patterns (First and Hollibaugh 2010). Appropriate software (e.g. ZooImage; FlowCam) with intelligent learning algorithms will in the future allow for an increasingly automatized and refined digital identification, enumeration, sorting and classification of meiofauna. In this way, we may even counteract the shortage of taxonomic experts, one of the most severe obstacles and disadvantages of working with meiofauna (see Lindgren et al. 2013).

Compared to traditional methods, molecular screening (also termed metabarcoding, second-generation sequencing, ultra- or meta-sequencing, high-throughput sequencing, environmental DNA monitoring), in principle, allows for an increasingly detailed examination of complete meiofauna samples at reduced time and personnel efforts, regardless whether applied for taxonomic scrutiny, diversity assessment, impact analysis or biogeographical comparison (Fonseca et al. 2010; 2017; Brannock et al. 2018). Through this growing sampling efficiency, meiofauna can assume their powerful role in impact studies urgently needed in the future (e.g. deep-sea mining).

Although gene-based analyses will dominate future meiobenthos research (see Sect. 1 in Chap. 4), there still remain open questions in their evaluation. Why are analyses of genetic fingerprints in some taxa, e.g. nematodes, so problematic, while annelids or harpacticoids are more accessible in this respect? What are the reasons for the difficult matching of sequenced data with morphological/structural data? While technical efforts and costs will further decrease, there remain methodological challenges and interpretational pitfalls (Brannock and Halanych 2015; Creer et al. 2015): choice of the most informative genetic provenance (mitochondrial, nuclear, somatic) and appropriate gene region, number of loci, representative sampling, quality of computer programs.

Choice of the right genetic marker seems to have a major influence on diversity estimates: in a comparative evaluation of an extensive set of sequences, Tang et al. (2012) could show for meiofauna that surveys based on the widely used nuclear 18S eDNA tended to severely underestimate diversity. They recommend instead the use of the cytochrome C oxidase (COI) subunit in the mitochondrial DNA (but see contrasting results in Avó et al. (2017) that rather speak for 18S rRNA, or see Pereira et al. (2010), who, for nematodes, point to a superiority of 28S sequences). Apparently, general procedures applicable for all groups and purposes are not advisable. But considering the confusing variety of methods applied in the rich literature, a broad standardization seems urgently needed to achieve robust and better comparable results. Guidelines for a DNA-based meiofauna taxonomy have been compiled by Fontaneto et al. (2015), and methodological recommendations for an 'environmental' DNA monitoring (albeit focussing on plankton samples in ponds) are well summarized (Fig. 1) by Harper et al. (2019).

Misinterpretations can be avoided when analyses are based on a broader genetic background instead of a few selected gene sequences only. Increasing computer power enables large-scale analyses, e.g. of the mitogenome, the entire genome, even of proteomes and metabolomes. These holistic analyses, already widespread in microbiology (see De Bruijn 2011), will increase and find broad applications also in meiobenthology.

As an alternative to DNA taxonomy and without the need of specific primers, analyses of proteome fingerprints on the basis of a novel mass spectrometric method (MALDI-TOF MS) have been performed for rapid identification of microorganisms as well as larger metazoans. Rossel and Arbizu (2018a, b) successfully introduced this method in meiobenthology for identifying small meiobenthic harpacticoids. They could reach a high level of accuracy by combining the cluster data with the supervised self-learning computer program Random Forest (RF) and introduce their work as a 'reliable method in biodiversity assessment' that outcompetes 'time-consuming and expensive barcoding'. If certain restrictions in storage of samples are observed, this method even works well with fixed samples—a further step towards automatic species identification with an unimagined potential.

Despite the improving technical support and rising experience, the valid interpretation of molecular results remains depending on the extent and quality of reference databases stored in databanks and inventories. A reliable documentation of molecular operational taxonomic units (OTUs) in large databanks, with detailed accompanying information on methods and provenience, is just one side of the coin. Why? All sequencing approaches find their limits in the rigour of taxa delimitation and

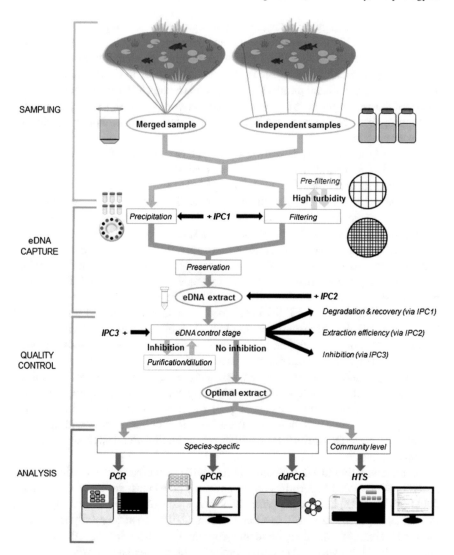

Fig. 1 Workflow for assessment of eDNA in water of a pond (from: Harper et al. 2019)

definition. Hence, comparison of sequences with their taxonomic counterparts, e.g. morphospecies, as reference bases, stored in museum collections worldwide, remains indispensable. In this context, a most relevant methodological advancement would be the possibility of extracting DNA from archived museum material, as suggested by Bhadury et al. (2007). If proven widely applicable, this progress would bridge many gaps between traditional and 'modern' systematic approaches.

The best-known and probably most comprehensive data inventory in the field of meiobenthology is WoRMS (World Register of Marine Species), an open-access databank that also integrates the former nematode database NeMyS). Beyond that, diverse special electronic libraries cover special areas or particular taxa: researchers on subterranean meiofauna find a separate centre in WoRCS (World Register of marine Cave Species), foraminiferans are covered by the WORLD FORAMINIFERA DATABASE, a part of WoRMS, and North Sea harpacticoids are registered in a library currently established in Germany. Promoting the GBOL inventory (German Barcode Of Life) it is planned to establish a united taxonomic and electronic reference centre of the countries' complete flora and fauna with an international and standardized access (Geiger et al. 2016). All of these inventories or registers, many of them as yet incomplete, have to be cross-validated and electronically interconnected—a huge task, but a necessity for future effective work in all meiobenthic fields, non-applied or applied.

Finally, perhaps a philosophical and subjective aspect: there is one serious disqualification in a purely molecular-based taxonomy: pursuant to the properties of human cognition, it is hard to imagine that molecular sequences and not their counterpart, the real animal, could find societal/political attention and protection. Hence, classical taxonomy will retain its place even in the future. All future taxonomic specialists should realize that.

2 New Trends Reviving 'Old Morphology'

A revolution: perfect morphological reconstructions – Neuroanatomy of meiobenthic groups reveals evolutionary old branches – Analyses down to the cellular basis by microscopical and histological methods – Towards ultrastructure inspections under non-destructive conditions – Future morphological meiobenthos studies nearer to focus of attention?

Whoever admired the brilliant fluorescence of selectively stained nerve structures in a confocal laser-scanning microscope (Fig. 2) or marvelled at three-dimensional computer reconstructions of organs based on serial sections (Fig. 3) will agree that in the last decades, morphology experienced a revolution. It is this new platform with often perfectly analysed details that allow for new histological and phylogenetic conclusions, hardly imaginable in earlier decades. Synchroton X-ray or phase-contrast tomography and holographic depiction of structural details (holographic field microscopy) are no longer science fiction or future illusions. They allow for 3D views of structures in an unprecedented size range down to chromosomal details. Raman microspectrometry can be combined and supplemented by SEM or TEM ultrastructural analysis (Maurin et al. 2010). Even specific histochemical patterns and biochemical pathways can be visualized by NanoSIMS microscopy (Jiang et al. 2016). Combining and correlating various new techniques seems to be the clue for an amazing progress in biological studies over the last decade, often resulting in central arguments for new phylogenetic paradigms (see Sect. 3): Using staining patterns of tissues that react selectively to specific antisera and neurotransmitters, Mayer et al. (2013) investigated one of the disputed phylogenetical furcations in the animal tree,

(a)   (b)   (c)   (d)

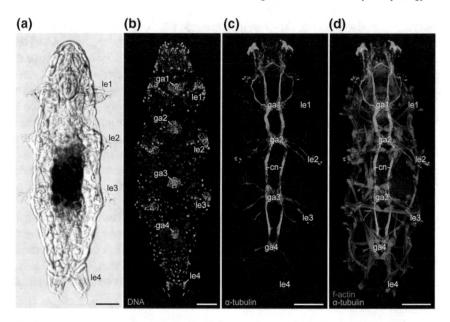

Fig. 2 Neuronal structures in various tardigrades visualized in confocal laser-scanning microscope after labelling with various immunoreactive markers; scale bars 25 μm (from: Mayer et al. 2013)

the position of Tardigrada within the Panarthropoda. Their histological results support a closer relationship of tardigrades to arthropods than to onychophorans and question relations to nematodes. Another example of a beautiful visualization and informative 3D microanatomy alike: Brenzinger et al. (2013) could not only clarify anatomical details of worm-like meiobenthic molluscs, but also gave important information on general evolutionary pathways in meiobenthos where convergences through paedomorphosis and miniaturization (the 'meiofauna syndrome') often tend to blur the picture.

For debates on kinship relations, structural details of the relatively 'conservative' brain and nervous systems and the neuromuscular patterns are of central relevance. Neural diversity with their systematic significance can give fundamental information about ancestral branching patterns. Some examples: the neuro-architecture of Gnathostomulida resembles conditions in Micrognathozoa underlining the basal position of Gnathifera and pointing to an ancestral, plesiomorphic status in Spiralia (Bekkouche and Worsaae 2016; Gąsiorowski et al. 2017). Neuroanatomical details in minute dinophilid worms revealed a fascinating correlation between brain regions and genetic expression patterns (Kerbl et al. 2016). On the basis of this conformance, general trends in the neural anatomy, conserved even in distantly related annelids, were disclosed. Rather early has the entire body of the biological model organism, the nematode *Caenorhabditis elegans,* been analysed down to the cellular level, and the complete structure of its nervous system could be deduced from serial sections (White et al. 1986). Today, progress in methodology (immuno-histochemistry and

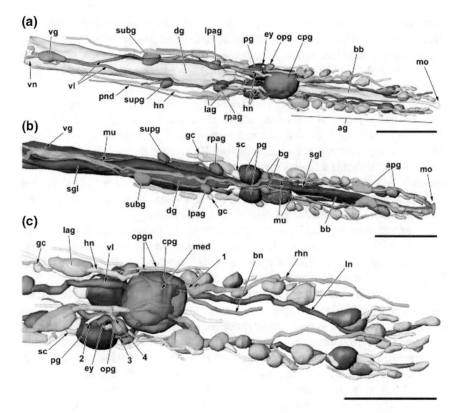

Fig. 3 *Helminthope psammobionta* (Mollusca), anterior end, scale bars 100 μm. 3D-computer reconstructions of internal organs evaluating serial sections (from: Brenzinger et al. 2013)

confocal laser-scanning microscopy) enables easier access to graphically expressive data. Yet, only few morphologists, so it seems, take this path that is applicable not only to morphological and phylogenetic deductions, but also to functional analyses.

Not only internal organs, also external hard structures such as the complex layering of chitinous components can be visualized by recent technical developments. Laser-scanning or epifluorescence microscopy helps analysing anatomical details for taxonomic and phylogenetic clarification. In studies on meiobenthic taxa with complex cuticles or teeth (e.g. in Scalidophora), these scientific methods call for wide application. Studying the composition of the feeding apparatus in tardigrades, Savic et al. (2016) found minute differences in the chitin composition that corresponded to local mechanical requirements and allowed for conclusions on food uptake. Using fluorescence effects, textural details of tardigrade plates became visible that had relevance for taxonomic conclusions (Perry et al. 2015).

Commenting these recently developed microscopical techniques is not to leave classical electron microscopy unmentioned. As shown in a study on aberrant polychaetes (Purschke and Jördens 2007), ultrastructural analyses remain valuable tools for assessing taxonomic relationships between sister groups and scrutinizing the evolutionary nature of structural similarities. Combined with serial sectioning and computer-based evaluation, 3D ultrastructure, although labour-intensive, could become crucial for reconstruction and visualization of complex biological components such as neuronal networks.

For those working on ultrastructural details with relevance for ecological problems, the development of 'environmental electron microscopy' (ESEM and ETEM) that allows inspection of 'wet', not denatured, biological material is certainly a most promising development. Combined with 3D imaging techniques (see Denk and Horstmann 2004), this technique could open unimagined possibilities for meiofauna research. Although introduced in chemical and technical fields already at the end of the last century, ESEM studies, applied to meiobenthic organisms, could not yet be found among the available publications. In microbiology, already today linking stable-isotope probing with Raman microspectrometry allows for 'fingerprinting' on a cellular level in a label-free and non-destructive manner. This enables a novel functional and ecological understanding in microbial communities (Wang et al. 2016). Why not also in the future world of meiobenthos?

Computer-assisted microscopy, both on micro- and ultrastructural levels, will not only add new morphological aspects to meiobenthology. The recent increase in novel methodologies will break new ground in our understanding of general functional and phylogenetic interrelations. Thus, by new techniques, the role of meiofauna can be better assessed and receive a higher attention in the general benthic realm. Surprising evolutionary pathways and connections of meiofauna with macrofauna can be revealed (see Sect. 3). Confirming these conclusions, many of the meiofauna publications cited before are published in highly regarded biological journals, not in specialized contributions noticed by a few readers only. This renders meiofaunal morphology and taxonomy an important future perspective and could give meiobenthology a central place in the understanding of modern zoology.

3   Phylogeny and Evolution—Meiofauna at the Deep Nodes

Phylogenomic trees increasingly stable – Novel 'millimetric' fossils of meiofauna – Metazoan origin in the Precambrian with meiofauna in key position – Some (discussed) phylogenetic relationships – Comprehensive keys by globally connected research

The desire to discover the origins, to see the archaic interrelations, which became obscured through millions of years, seems inherent to mankind. Therefore, clarification of evolutionary trends and reconstruction of evolutionary patterns are among the central goals of biological thinking. Today, new sophisticated methods are applied in phylogenetic reconstructions of 'deep branching' relationships. And they often reveal the central position of many meiobenthic taxa. 'Small animals provide clues to large-scale questions' as K. Worsaae from Denmark had stated (ISIMCO 2016).

These ancestors of minute size bring meiofauna studies more and more in the focus of general interest.

These days, multiple and comprehensive genomic and transcriptomic analyses unravel phylogenetic relationships, similarities or discrepancies between taxa, that, some decades ago, were completely unimaginable to biologists (e.g. the taxon Ecdysozoa). Thus, many molecular-based phylogenetic deductions have refuted traditional morphology-based trees of life. The increasing availability of genome sequences allows for detailed gene exploration and puts evolutionary trees on a broad phylogenomic basis with considerably reduced sources of error. In many cases, the new approaches have shown that earlier phylogenetic conclusions deducted from a few 'model animals' were prone to fail. On the other hand, the new computer-based phylogenomic patterns often receive support by modern morphological evidence (s. Sect. 2).

Moreover and most importantly, tree constructions get increasing validation by the fossil record. Serendipitous discoveries of new *lagerstätten* with rich 'millimetric fossils' and even well-discernable submillimetric internal structures are becoming more frequent. Their often exquisite preservation and sophisticated analysis (e.g. non-contact X-ray micro-tomography) allow for calculations of time-trees that probably push back the origin of first meiobenthic lineages deep into the Early Cambrian, probably even the Late Ediacaran (Rota-Stabelli et al. 2013; Yang et al. 2016; Han et al. 2017; Harvey and Butterfield 2017; Zhang and Xiao 2017). Among these early beings, remnants of tiny Loriciferans with only 300 μm in length have been found. Using microtomography, Parry et al. (2017) found 'meiofaunal tracemakers' (ichnofossils), probably of nematoid origin, from the Late Ediacaran period in Brazilian siltstones. The parallel trackways of minute traces (footprints) recently discovered in China (Chen et al. 2018), were they produced by the appendages of bilateral meiobenthic annelids or arthropods that lived about 550 million years ago?

These archaic meiofauna remnants, often even three-dimensionally phosphatized (*Orsten*-type), show the early origin of a benthic (interstitial?) lifestyle, but also document meiofauna-type animals as first colonizers of benthic layers in the Early Cambrian or even Late Ediacaran oceans. This would refute the supposition that the ocean floor in the Early Cambrian was 'monopolized by protists' (Harvey and Butterfield 2017). Now, meiofaunal evolution seems to reach back to hitherto unexplored, archaic time periods: meiofauna as the earliest infauna. Both molecular and fossil evidence underline the possibility that animals of meiofaunal size have key positions in the Bilateria, both in the protostome and in the deuterostome trees (Fig. 4, see Laumer et al. 2015; Cunningham et al. 2017; Han et al. 2017; Chen et al. 2018). These findings challenge the disputed 'Cambrian explosion' which, according to Cunningham et al. (2017), 'was neither Cambrian nor explosive', and refute problems with 'Darwin's Dilemma' (the 'abrupt' appearance of taxa with novel body structures).

No rule without exception—in annelids, the developmental trend from small to large might have been reversed: Struck et al. (2015) concluded from phylogenomic analyses that in this group, the larger representatives were first. According to their arguments, meiobenthic annelids are to be derived from larger ancestral forms, which

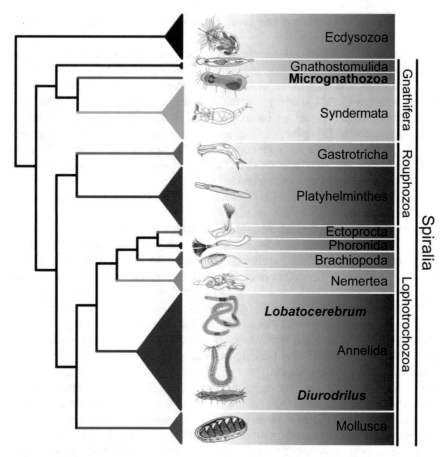

Fig. 4 Phylogenetic tree of major spiralian taxa displaying three main groups within Spiralia (from: Laumer et al. 2015, modified)

stepwise adapted to their interstitial environment by progenesis and miniaturization, developmental trends also found in psammodrilid annelids (Worsaae et al. 2018).

The phylogenetic approaches outlined above—reconstructions from fossil remains and molecular analyses—have their incongruities based on various pitfalls. In calculations based on genome analyses, significant divergences can occur in the rate of variations, since these mainly depend on correlations with generation time and can result in a potential bias when tuning the molecular clocks (Thomas et al. 2010). Bleidorn (2017) points to other possible systematic errors in phylogenomic studies. He refers to problems of taxon sampling and choice of 'correct' genes with high information content ('highly saturated genes should be avoided'). In addition, processes like extensive convergent evolution, observed to occur especially in nematodes, may exert an aggravating effect on the interpretation of phylogenetic trees and may possibly explain problems in their DNA-based phyletic reconstruction (Van Megen et al. 2009).

Even the fossil records presently available cannot represent 'a perfect archive of animal evolutionary history' (Donoghue, ISIMCO 2016, unpubl.), since the understanding of fossils and their timing by calibrating molecular clocks with the ambient geological setting become increasingly problematical over these vast periods of time. Although the use of sophisticated new methods will enhance validity of phyletic analyses (Cunningham et al. 2017), divergent interpretations of results may remain.

The following short compilation of some more general phylogenetic conclusions (sometimes debated suggestions!) underlines the crucial role of meiofaunal clades in the reconstruction of animal evolution. These frequently surprising cognitions document the considerable progress in research on animals of meiobenthic size. Regardless of possible future corrections, the basal position of meiofaunal representatives in the genealogical tree of most ancient clades will be supported.

- Xenacoelomorpha (Acoelomorpha + *Xenoturbella*) are considered a sister group to the remaining Bilateria (Nephrozoa) (Cannon et al. 2016)
- Scalidomorpha (Kinorhyncha, Priapulida, perhaps including Loricifera) are sister to remaining Ecdysozoa (Laumer et al. 2015)
- Early euarthropods show the nervous system of ancestral Ecdysozoa (Yang et al. 2016)
- Nematoda and Kinorhyncha are related (Park et al. 2006; Borner et al. 2014)
- Nematoda existed already in the Early Ordivician (470 Mio years) (Bali´nski et al. 2013); nematoid traces have been found in the Late Ediacaran (Parry et al. 2017); traces of parallel appendages (arthropods or annelids?) date back to the terminal Ediacaran in China (Chen et al. 2018)
- Gnathifera (Rotifera, Gnathostomulida, Micrognathozoa) are a monophyletic group and sister to Spiralia (Laumer et al. 2015; Gąsiorowski et al. 2017)
- Spiralia are monophyletic and have most probably a meiofaunal-sized, unsegmented ancestor (Witek et al. 2009; Laumer et al. 2015).
- Spiralia perhaps include the monophyletic taxon Lophotrochozoa (Witek et al. 2009)
- Deuterostomes had ancestors of meiofaunal size with lateral gills (Han et al. 2017)
- Ostracod crustaceans are of pre-Silurian age (Siveter et al. 2014), 'bivalved ecdysozoa' of meiofaunal size are to be found in unusual richness already in the early Cambrian/terminal Ediacaran (Zhang and Xiao 2017).

Supported by computational progress, the integrated efforts of molecular biologists and morphology-based taxonomists create worldwide collaborations and, based on reference libraries, contribute to a wealth of information that allows for an unprecedented increase in meiobenthic knowledge. Who could imagine a complete three-dimensional myoanatomical reconstruction (Neves et al. 2013)? In the field of morphology/taxonomy, identification keys are not any longer restricted to some genera or families. They now cover entire groups of high taxonomic rank or wide, even world-wide geographical range (Fontaneto et al. 2007). In the earlier days of meiobenthology, who would have dreamed of global treatments on freshwater nematodes (Abebe et al. 2008) or groundwater copepods (Galassi et al. 2009) or halacarids (Bartsch 2009)?

In recent years, progress in molecular analyses, in digitally supported microscopic instrumentation, in refined processing of paleontological relics, produced an unprecedented wealth of genetic, anatomical and paleontological data on meiobenthos that gave morphological analysis and taxonomic ordination a fascinating and directive drive. This new role animals of meiobenthic size attain in phylogenetic considerations not only reveals new phyletic relations and refutes many of the traditional conceptions. This process will shift meiofauna from a marginal position into the focus of experts' studies and warrant generally relevant front-page publications. This is betrayed by many of the journals cited in this volume, and there is no reason to assume that this trend will come to an end.

References

Abebe E, Decraemer W, Ley PD (2008) Global diversity of nematodes (Nematoda) in freshwater. Hydrobiologia 595:67–78

Avó AP, Daniell TJ, Neilson R et al (2017) DNA barcoding and morphological identification of benthic nematodes assemblages of estuarine intertidal sediment: advances in molecular tools for biodiversity assessment. Front Mar Sci 4:66. https://doi.org/10.3389/fmars.2017.00066

Bali´nski A, Sun Y, Dzik J (2013) Traces of marine nematodes from 470 million years old early Ordovician rocks in China. Nematology 15:567–574

Bartsch I (2009) Checklist of marine and freshwater halacarid mite genera and species (Halacaridae: Acari) with notes on synonyms, habitats, distribution and descriptions of the taxa. Zootaxa 1998:3–170

Bekkouche N, Worsaae K (2016) Nervous system and ciliary structures of Micrognathozoa (Gnathifera): evolutionary insight from an early branch in Spiralia. R Soc Open Sci 3(10):160289

Bhadury PMC, Austen DT, Bilton PJD et al (2007) Exploitation of archived marine nematodes—a hot lysis DNA extraction protocol for molecular studies. Zool Scr 36:93–98

Bleidorn C (2017) Sources of error and incongruence in phylogenomic analyses. Phylogenomics—An introduction. Springer, Berlin, pp 173–193

Borner J, Rehm P, Schill RO et al (2014) A transcriptome approach to ecdysozoan phylogeny. Mol Phylogen Evol 80:79–87

Brannock PM, Halanych KM (2015) Meiofaunal community analysis by high-throughput sequencing: comparison of extraction, quality filtering, and clustering methods. Mar Genomics 23:67–75

Brannock PM, Learman DR, Mahon AR et al (2018) Meiobenthic community composition and biodiversity along a 5500 km transect of Western Antarctica: a metabarcoding analysis. Mar Ecol Prog Ser 603:47–60. https://doi.org/10.3354/meps12717

Brenzinger B, Haszprunar G, Schrödl M (2013) At the limits of a successful body plan—3D microanatomy, histology and evolution of *Helminthope* (Mollusca: Heterobranchia: Rhodopemorpha), the most worm-like gastropod. Front Zool 10:37. www.frontiersinzoology.com/content/10/1/37

Cannon JT, Vellutini BC, Smith J et al (2016) Xenacoelomorpha is the sister group to Nephrozoa. Nature 530:89–93

Chen Z, Chen X, Chuanming Z et al (2018) Late Ediacaran trackways produced by bilaterian animals with paired appendages. Sci Adv 4:eaao6691

Costello MJ (2015) Biodiversity: the known, unknown, and rates of extinction. Curr Biol 25(9):R368–R371. https://doi.org/10.1016/j.cub.2015.03.051

Creer S, Fonseca VG, Porazinska DL et al (2015) Ultrasequencing of the meiofaunal biosphere: practice, pitfalls and promises. Mol Ecol 19(Suppl. 1):4–20

Cunningham JA, Liu AG, Bengtson S, Donoghue PCJ (2017) The origin of animals: can molecular clocks and the fossil record be reconciled? BioEssays 39:1600120. https://doi.org/10.1002/bies.201600120

De Bruijn FJ (ed) (2011) Handbook of molecular microbial ecology II. Metagenomics in different habitats. Wiley-Blackwell, Wiley & Sons Inc, Hoboken, New Jersey, 610 pp

Denk W, Horstmann H (2004) Serial block-face scanning electron microscopy to reconstruct three-dimensional tissue nanostructure. PLoS Biol 2(11):e329. https://doi.org/10.1371/journal.pbio.0020329

Derycke S, De Ley P, De Ley IT et al (2010) Linking DNA sequences to morphology: cryptic diversity and population genetic structure in the marine nematode *Thoracostoma trachygaster* (Nematoda, Leptosomatidae). Zool Scr 39:276–289

First MR, Hollibaugh JT (2010) Diel depth distributions of microbenthos in tidal creek sediments: high resolution mapping in fluorescently labeled embedded cores. Hydrobiologia 655:149–158

Fonseca VG, Carvalho GR, Sung W et al (2010) Second-generation environmental sequencing unmasks marine metazoan biodiversity. Nat Commun 1:98. https://doi.org/10.1038/ncomms1095

Fonseca VG, Packer M, Carvalho GR et al (2011) Isolation of marine meiofauna from sandy sediments: from decanting to DNA extraction. Nat Protoc Exch. https://doi.org/10.1038/nprot.2010.157

Fonseca VG, Sinniger F, Gaspar JM et al (2017) Revealing higher than expected meiofaunal diversity in Antarctic sediments: a metabarcoding approach. Sci Rep 7:6094. https://doi.org/10.1038/s41598-017-06687-x

Fontaneto D, De Smet WH, Melone G (2007) Identification key to the genera of marine rotifers worldwide. Meiofauna Mar 16:75–99

Fontaneto D, Flot J-F, Tang CQ (2015) Guidelines for DNA taxonomy, with a focus on the meiofauna. Mar Biodivers 45:433–451

Galassi DMP, Huys R, Reid JW (2009) Diversity, ecology and evolution of groundwater copepods. Freshw Biol 54:691–708

Gąsiorowski L, Bekkouche N, Worsaae K (2017) Morphology and evolution of the nervous system in Gnathostomulida (Gnathifera, Spiralia). Org Divers Evol 17:447. https://doi.org/10.1007/s13127-017-0324-8

Geiger M, Borsch J, Burkhardt U (2016) How to tackle the molecular species inventory for an industrialized nation—lessons from the first phase of the German Barcode of Life initiative (GBOL (2012–2015). Genome 59:1–10

Grosemans T, Morris K, Thomas WK et al (2016) Mitogenomics reveals high synteny and long evolutionary histories of sympatric cryptic nematode species. Ecol Evol 6:1854–1870. https://doi.org/10.1002/ece3

Han J, Conway Morris S, Ou Q et al (2017) Meiofaunal deuterostomes from the basal Cambria of Shaanxi (China). Nature 542:228–231

Harper LR, Buxton AS, Rees HC et al (2019) Prospects and challenges of environmental DNA (eDNA) monitoring in freshwater ponds. Hydrobiologia 826:25–41. https://doi.org/10.1007/s10750-018-3750-5

Harvey THP, Butterfield NJ (2017) Exceptionally preserved Cambrian loriciferans and the early animal invasion of the meiobenthos. Nat Ecol Evol 1:0022. https://doi.org/10.1038/s41559-016-0022

ISIMCO (2016) Draft book of abstracts. In: 16th international meiofauna conference, Heraklion, Greece, 3–8 July 2016, 182 pp

Jiang H, Kilburn MR, Decelle J, Musat N (2016) NanoSIMS chemical imaging combined with correlative microscopy for biological sample analysis. Curr Opin Biotechnol 41:130–135

Karanovic T, Kim K (2014) New insights into polyphyly of the harpacticoid genus *Delavalia* (Crustacea, Copepoda) through morphological and molecular study of an unprecedented diversity of sympatric species in a small South Korean bay. Zootaxa 3783:1–96. https://doi.org/10.11646/zootaxa.3783.1.1

Kerbl A, Martin-Durán JM, Worsaae K, Hejnol A (2016) Molecular regionalization in the compact brain of the meiofauna annelid *Dinophilus gyrociliatus* (Dinophilidae). EvoDevo 7:20

Kitahashi T, Watanabe HK, Tsuchiya M et al (2018) A new method for acquiring images of meiobenthic images using the FlowCAM. MethodsX 5:1330–1335. https://doi.org/10.1016/j.mex.2018.10.012

Laumer CE, Bekkouche N, Kerbl A et al (2015) Spiralian phylogeny informs the evolution of microscopic lineages. Curr Biol 25:2000–2006

Lindgren JF, Hassellöv I-M, Dahllöf I (2013) Analyzing changes in sediment meiofauna communities using the image analysis software ZooImage. J Exp Mar Biol Ecol 440:74–80

Maurin LC, Himmel D, Mansot JL et al (2010) Raman microspectrometry as a powerful tool for a quick screening of thiotrophy: an application on mangrove swamp meiofauna of Guadeloupe (FWI). Mar Environ Res 69:382–389

Mayer G, Martin C, Rüdiger J et al (2013) Selective neuronal staining in tardigrades and onychophorans provides insights into the evolution of segmental ganglia in panarthropods. BMC Evol Biol 13:230. https://doi.org/10.1186/1471-2148-13-230

Neves R, Bailly X, Leasi F et al (2013) A complete three-dimensional reconstruction of the myoanatomy of Loricifera: comparative morphology of an adult and a Higgins larva stage. Front Zool 10:1–21

Park JK, Rho HS, Kristensen RM et al (2006) First molecular data on the phylum Loricifera—an investigation into the phylogeny of Ecdysozoa with emphasis on the positions of Loricifera and Priapulida. Zool Sci 23:943–954

Parry LA, Boggiani PC, Condon DJ et al (2017) Ichnological evidence for meiofaunal bilaterians from the terminal Ediacaran and earliest Cambrian of Brazil. Nature Ecol Evol 1:1455–1464. https://doi.org/10.1038/s41559-017-0301-9

Pereira TJ, Fonseca G, Mundo-Ocampo M et al (2010) Diversity of free-living marine nematodes (Enoplida) from Baja California assessed by integrative taxonomy. Mar Biol 157:1665–1678. https://doi.org/10.1007/s00227-010-1439-z

Perry ES, Miller WR, Lindsay S (2015) Looking at tardigrades in a new light: using epifluorescence to interpret structure. J Microsc 257(2):117–122

Purschke G, Jördens J (2007) Male genital organs in the eulittoral meiofaunal polychaete *Stygocapitella subterranea* (Annelida, Parergodrilidae): ultrastructure, functional and phylogenetic significance. Zoomorphology 126:283–297

Rossel S, Martinez Arbizu PM (2018a) Automatic specimen identification of harpacticoids (Crustacea: Copepoda) using random forest and MALDI-TOF mass spectra, including a post hoc test for false positive discovery. Methods Ecol Evol 2018:1–14. https://doi.org/10.1111/2041-210X.13000

Rossel S, Martinez Arbizu PM (2018b) Effects of sample fixation on specimen identification in biodiversity assemblies based on proteomic data (MALDI-TOF). Front Mar Sci 5:149. https://doi.org/10.3389/fmars.2018.00149

Rota-Stabelli O, Daley AC, Pisani D (2013) Molecular timetrees reveal a Cambrian colonization of land and a new scenario for Ecdysozoan evolution. Curr Biol 23:392–398

Savic AG, Preus S, Rebecchi L, Guidetti R (2016) New multivariate image analysis method for detection of differences in chemical and structural composition of chitin structures in tardigrade feeding apparatuses. Zoomorphology 135:43–50

Siveter DJ, Tanaka G, Farrell UC et al (2014) Exceptionally preserved 450-million-year-old Ordovician ostracods with brood care. Curr Biol 24:801–806

Struck TH, Golombek A, Weigert A et al (2015) The evolution of annelids reveals two adaptive routes to the interstitial realm. Curr Biol 25:1933–1999

Tang CQ, Leasi F, Obertegger U et al (2012) The widely used small subunit 18S rDNA molecule greatly underestimates true diversity in biodiversity surveys of the meiofauna. PNAS 109:16208–16212

Thomas JA, Welch JJ, Lanfear R, Bromham L (2010) A generation time effect on the rate of molecular evolution in invertebrates. Mol Biol Evol 27:1173–1180

Van Megen H, Van Den Elsen S, Holterman M et al (2009) A phylogenetic tree of nematodes based on about 1200 full-length small subunit ribosomal DNA sequences. Nematology 11:927–950

Wang Y, Huang WE, Cui L, Wagner M (2016) Single cell stable isotope probing in microbiology using Raman microspectroscopy. Curr Opin Biotechnol 41:34–42

White JG, Southgate E, Thomson JN, Brenner S (1986) The structure of the nervous system of the nematode *Caenorhabditis elegans*. Phil Trans R Soc Lond B 314:1–340

Witek A, Herlyn H, Ebersberger I et al (2009) Support for the monophyletic origin of Gnathifera from phylogenomics. Mol Phylogen Evol 53:1037–1041

Worsaae K, Giribet G, Martínez A (2018) The role of progenesis in the diversification of the interstitial annelid lineage Psammodrilidae. Inv Sys 32:774

Xu J, Wang Y-S, Yin J, Lin J (2011) New series of corers for taking undisturbed vertical samples of soft bottom sediments. Mar Environ Res 71:312–316

Yang J, Ortega-Hernández J, Butterfield NJ et al (2016) Fuxianhuiid ventral nerve cord and early nervous system evolution in Panarthropoda. Proc Nat Acad Sci 113:2988–2993

Zhang H, Xiao S (2017) Three-dimensionally phosphatized meiofaunal bivalved arthropods from the Upper Cambrian of Western Hunan, South China. Neues Jb Geol Paläont—Abhdl 285:39–52

Chapter 7
Epilogue

This treatise accentuates future directions that should lead meiobenthology out of a frequently disregarded niche into mainstream benthology and zoology. Varying some lines by Fonseca et al. (2018), my appeal for a successful future of meiobenthology is:

- Address topics that are relevant also for general biological aspects.
- Coordinate and connect the often separated thematic lines/disciplines and methods by cooperations.
- Use the advantages of new technologies.
- Feed your own data into the pools of information collected in worldwide accessible data libraries.

These tasks may be complex and challenging, but they are the ones, so is my memorandum, that can bring meiobenthology nearer to the main road of benthic science. They will show to the broad scientific community and decision-makers alike that the lack of attention, meiofauna has often received, is no longer justified. These days neither methodological nor analytical and ecological constraints exist to ignore the advantages inherent in meiofauna research (Schratzberger and Ingels 2018).

My considerations on the present and future meiofauna research situation have been based on that part of meiofauna literature, which I considered contributing to general benthology and to our knowledge on processes that will gain general future attention. To overcome the risk of marginalization, we need enduring cooperative efforts from coordinated research groups, and we need to work with standardized methods at adjusted aims (Bellard et al. 2012). Opening up our little research nut-shells, connecting our goals with general trends in benthology, also 'conducting large meta-studies', would blow 'fresh wind' into the scene and even bring the new money needed for more complex and integrated research.

Working on this text, a critical 'bottleneck' situation in meiofauna research became apparent to me: most papers of notable relevance in the various fields covered here originate from some few, large and well-established 'research schools'. It seems that their future success will be decisive for promoting meiobenthology towards a more general influence and acceptance. In reverse conclusion, this implies that a potential decline of these centres, for whatever reasons, could easily bring meiobenthology in dire times. Against this background, future meiobenthological research should avoid all steps that might deepen the problematic split between marine and freshwater meiobenthology. We should rather try to unify these historically diverging fields (Giere 2009; Warwick 2018).

For future success, meiobenthology must evolve. Mastering the forthcoming developments and risks will ask for changing research strategies, far-reaching contacts and shared responsibilities. Certainly, these pathways are made possible and often even defined by progress in technologies. They will provide us with lots of increasingly automatized and computer-assisted data. Hence, the challenge of the future will not be gathering new data, it will rather lie in interpreting them. Creating new results is much easier than analysing their meaning and finding their right position in the huge body of knowledge accumulated through time. And this requires our human judgement. So, all future visions of meiobenthic science will depend on our personalities and on our experience and collaboration within the meiobenthos research community. Progress depends on us.

References

Bellard C, Bertelsmeier C, Leadley P et al (2012) Impacts of climate change on the future of biodiversity. Ecol Lett 15:365–377

Fonseca G, Fontaneto D, Di Domenico M (2018) Addressing biodiversity shortfalls in meiofauna. J Exp Mar Biol Ecol 502:26–38

Giere O (2009) Meiobenthology. The microscopic motile fauna of aquatic sediments, 2nd edn. Springer, Berlin, Heidelberg, pp 527

Schratzberger M, Ingels J (2018) Meiofauna matters: the roles of meiofauna in benthic ecosystems. J Exp Mar Biol Ecol 502:12–25. https://doi.org/10.1016/j.jembe.2017.01.007

Warwick R (2018) The contrasting histories of marine and freshwater meiobenthic research—a result of differing life histories and adaptive strategies? J Exp Mar Biol Ecol 502:4–11

Printed in the United States
By Bookmasters